Practical Introduction to Pumping Technology

Gulf Publishing Company
Houston, Texas

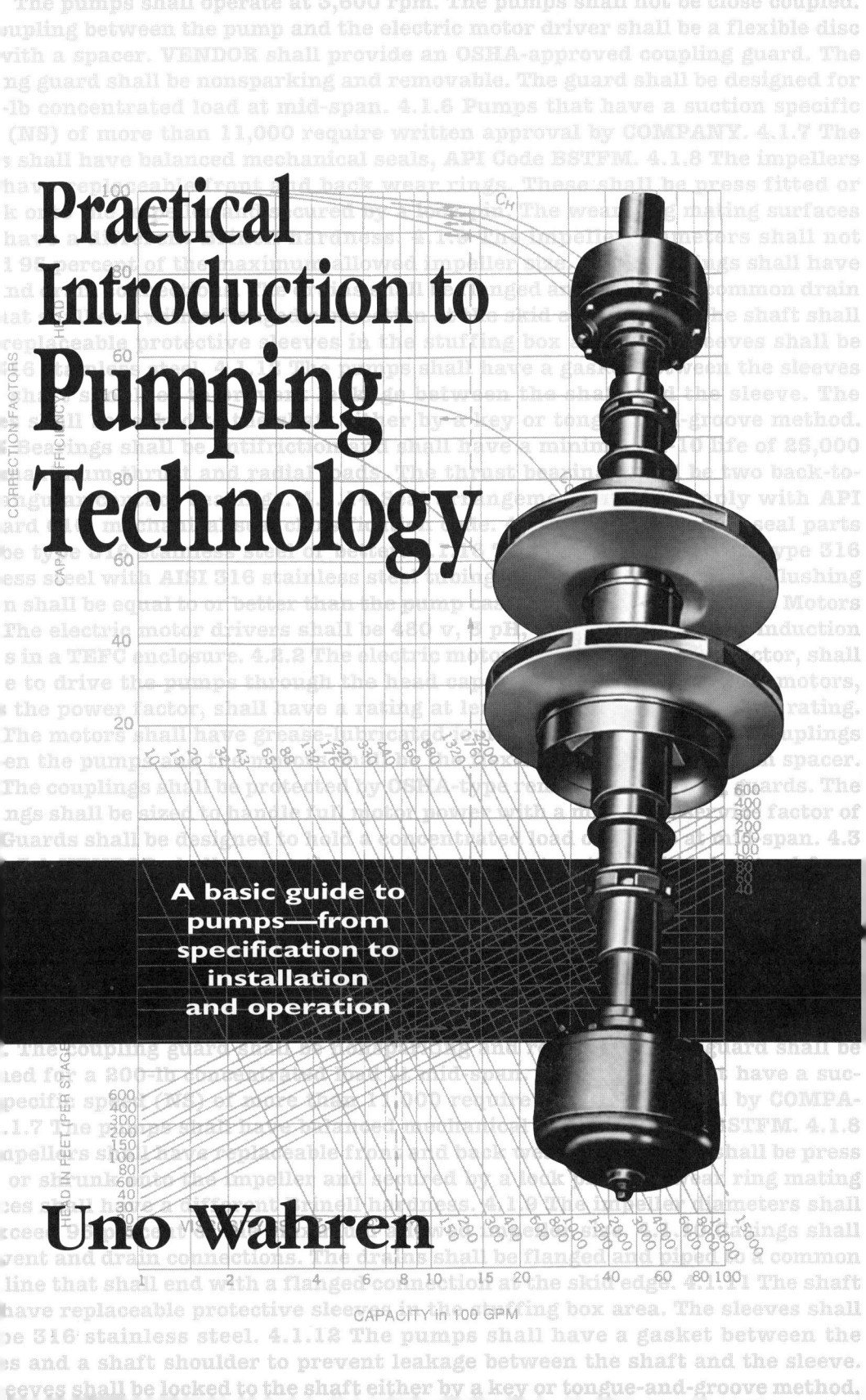

Practical Introduction to Pumping Technology

A basic guide to pumps—from specification to installation and operation

Uno Wahren

Practical Introduction to Pumping Technology

Copyright © 1997 by Gulf Publishing Company, Houston, Texas. All rights reserved. Printed in the United States of America. This book, or parts thereof, may not be reproduced in any form without permission of the publisher.

Gulf Publishing Company
Book Division
P.O. Box 2608 ☐ Houston, Texas 77252-2608

10 9 8 7 6 5 4 3 2

Library of Congress Cataloging-in-Publication Data

Wahren, Uno
    Practical introduction to pumping technology / Uno Wahren.
      p. cm.
    Includes bibliographical references and index.
    ISBN 0-88415-686-9 (alk. paper)
    1. Pumping machinery.   I. Title.
TJ900.W18   1997
621.6′9—dc21   97-18401
                                                                           CIP

Contents

Chapter 1
Parameters   1

Chapter 2
Pump Calculations   9
Friction, 9. Head Calculations, 10. Horsepower, 15. Specific Speed, 16. Suction Specific Speed, 17. Affinity Formulas, 17.

Chapter 3
Required Data for Specifying Pumps   19

Chapter 4
Pump Types   21
Centrifugal Pumps, 21. Axial-Flow and Mixed-Flow Pumps, 22. Radial-Flow Pumps, 22. Positive Displacement Pumps, 30. Reciprocating Pumps, 30. Rotary Pumps, 35. Special-Purpose Pumps, 39.

Chapter 5
Specifications   42
Data Sheets, 42. Specifications, 43.

Chapter 6
Pump Curves   45
Centrifugal Pump Curves, 45. Head Capacity Curves, 45. System Curves, 48. Pumps Operating in Parallel, 48. Pumps Operating in Series, 51. Positive Displacement Pump Curves, 54.

Chapter 7
Effects of Viscosity on Pump Performance   55
Dynamic (Absolute) Viscosity, 55. Kinematic Viscosity, 55. Viscosity Units, 55. Industry Preferences, 56.

Chapter 8
Vibration — 61
Terms and Definitions, 61. Testing Procedures, 62. Vibration Limits, 63. Induced Piping Vibrations, 65.

Chapter 9
Net Positive Suction Head (NPSH) — 66
Definition, 66. NPSH Calculations, 66. Additional Requirements, 71.

Chapter 10
Pump Shaft Sealing — 74
Packed Glands, 74. Mechanical Face Seals, 75. Cyclone Separator, 82. Flush and Quench Fluids, 82. Stuffing-Box Cooling, 82. Buffer Fluid Schemes, 82. Face Seal Life Expectancy, 82.

Chapter 11
Pump Bearings — 83
Bearing Types, 83. Bearing Lubrication, 89. Bearing Cooling, 91. Bearing Seals, 91.

Chapter 12
Metallurgy — 92
Corrosion, 92. Pump Materials, 93. Cast Iron, 93. Ferritic Steel, 93. Martensitic Stainless Steel, 97. Austenitic Stainless Steel, 97. Duplex Stainless Steel, 98. Nonferrous Materials, 98. Titanium, 99. Plastic, 99.

Chapter 13
Pump Drivers — 100
Electric Motors, 100. Internal Combustion Engines, 106. Steam Turbines, 109. Gas Turbines, 111. Hydraulic Drives, 113. Solar Power, 113.

Chapter 14
Gears — 114
Parallel Shaft Gears, 114. Right-Angle Gears, 118. Epicyclic Gears, 120.

Chapter 15
Couplings — 121
Types of Couplings, 121. Typical Service Factors, 127.

Chapter 16
Pump Controls — 128
Control Valve Types, 128. Capacity Control, 129. Minimum Flow Bypass, 132. Liquid Level Control, 132. On-Off Control, 133. Modulating Control, 133. Pressure Control, 133. Surge Control, 134. Control Selection for Positive Displacement Pumps, 134. Pulsation Dampeners, 136.

Chapter 17
Instrumentation — 137
Instruments, 137. Annunciators, Alarms, and Shutdowns, 137. Functions, 138. Electrical Area Classification, 139.

Chapter 18
Documentation — 140

Chapter 19
Inspection and Testing — 142
General Inspection, 142. Hydrostatic Test, 143. Performance Test, 143. NPSH Test, 145.

Chapter 20
Installation and Operation — 146
Installation, 146. Piping and Valves, 148. Pump Start-up, 149.

Chapter 21
Troubleshooting — 151
Centrifugal Pumps, 151. Reciprocating Pumps, 153.

Appendix 1
Sample Pump Specification — 154

Appendix 2
Centrifugal Pump Data Sheet — 160

Appendix 3
Internal Combustion Engine Data Sheet — 161

Appendix 4
Electric Motor Data Sheet — 162

Appendix 5
Centrifugal Pump Package   163

Appendix 6
Maximum Viable Suction Lifts at Various Altitudes   164

Appendix 7
Suggested List of Vendors   165

Appendix 8
***API-610* Mechanical Seal Classification Code**   175

References, 176

Index, 177

Chapter 1
Parameters

This book contains information needed to select the proper pump for a given application, create the necessary documentation, and choose vendors. Many books dealing with centrifugal and positive displacement pumps exist. Almost all these books cover pump design and application in great detail, and many are excellent. This author does not intend to compete head to head with the authors of these books, but to supply a compact guide that contains all the information a pump user or application engineer will need in one handy, uncomplicated reference book.

This book assumes the reader has some knowledge of hydraulics, pumps, and pumping systems. Because of space limitations, all hydraulic and material property tables cannot be included. However, excellent sources for hydraulic data include *Hydraulic Institute Complete Pump Standards* and *Hydraulic Institute Engineering Data Book*.

Hydraulics is the science of liquids, both static and flowing. To understand pumps and pump hydraulics, pump buyers need to be familiar with the following industry terminology.

Pressure

This term means a force applied to a surface. The measurements for pressure can be expressed as various functions of psi, or pounds per square inch, such as:

- Atmospheric pressure (psi) = 14.7 psia
- Metric atmosphere = psi × 0.07
- Kilograms per square centimeter (kg/cm^2) = psi × 0.07
- Kilopascals = psi × 6.89
- Bars = psi × 14.50

Atmospheric Pressure

The pressure exerted on a surface area by the weight of the atmosphere is atmospheric pressure, which at sea level is 14.7 psi, or one atmosphere. At higher altitudes, the atmospheric pressure decreases. At locations below sea level, the atmospheric pressure rises. (See Table 1.1.)

Table 1.1
Atmospheric Pressure at Some Altitudes

Altitude	Barometric Pressure	Equivalent Head	Maximum Practical Suction Lift (Water)
−1,000 ft	15.2 psi	35.2 ft	22 ft
Sea level	14.7 psi	34.0 ft	21 ft
1,500 ft	13.9 psi	32.2 ft	20 ft
3,000 ft	13.2 psi	30.5 ft	18 ft
5,000 ft	12.2 psi	28.3 ft	16 ft
7,000 ft	11.3 psi	26.2 ft	15 ft
8,000 ft	10.9 psi	25.2 ft	14 ft

Note: Water temperature = 75°F

Vacuum

Any pressure below atmospheric pressure is a partial vacuum. The expression for vacuum is in inches or millimeters of mercury (Hg). Full vacuum is at 30 in. Hg. To convert inches to millimeters multiply inches by 25.4.

Vapor Pressure

At a specific temperature and pressure, a liquid will boil. The point at which the liquid begins to boil is the liquid's vapor pressure point. The vapor pressure (vp) will vary with changes in either temperature or pressure, or both. Figure 1.1 shows the vapor pressure for propane as 10.55 psi at 60°F. At 120°F the vapor pressure for propane is 233.7 psi.

Gauge Pressure

As the name implies, pressure gauges show gauge pressure (psig), which is the pressure exerted on a surface minus the atmospheric pressure. Thus, if the absolute pressure in a pressure vessel is 150 psia, the pressure gauge will read 150 − 14.7, or 135.3 psig.

Absolute Pressure

This is the pressure of the atmosphere on a surface. At sea level, a pressure gauge with no external pressure added will read 0 psig. The atmospheric pressure is 14.7 psia. If the gauge reads 15 psig, the absolute pressure will be 15 + 14.7, or 29.7 psia.

Parameters

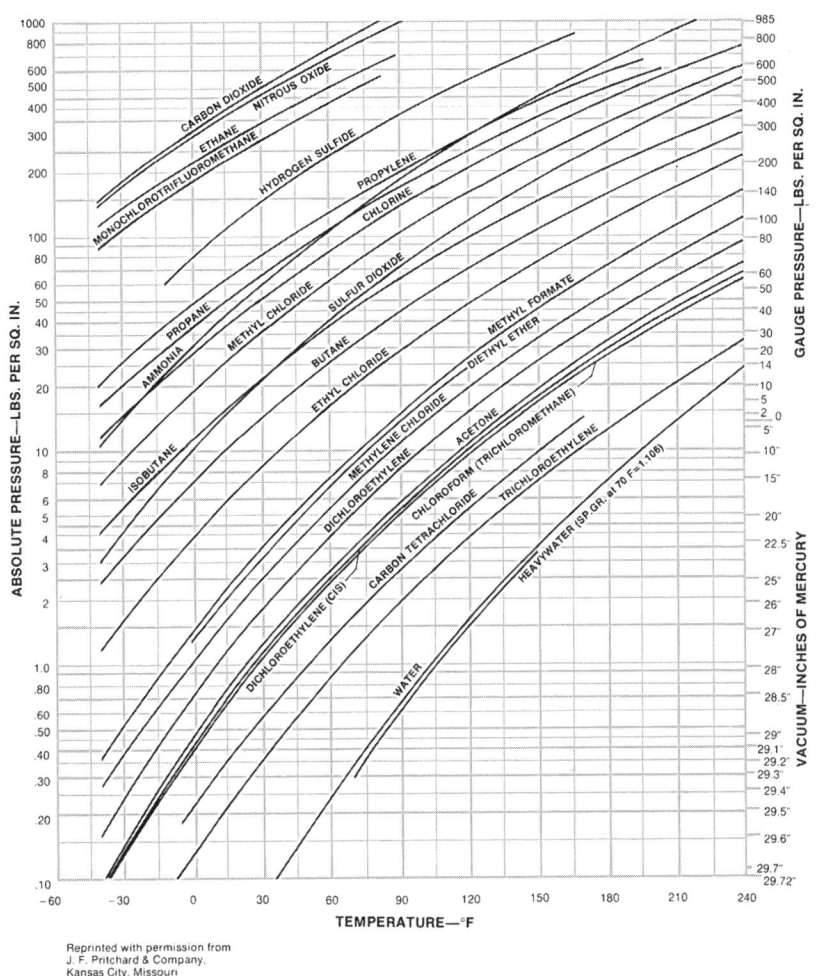

Figure 1.1 Vapor Pressure of Various Liquids, −60°F to 240°F (*Courtesy of the Hydraulic Institute*)

Flow

This term refers to the liquid that enters the pump's suction nozzle. Flow (Q) measurements are U.S. gallons per minute (USgpm or gpm) and can be converted as follows:

4 *Practical Introduction to Pumping Technology*

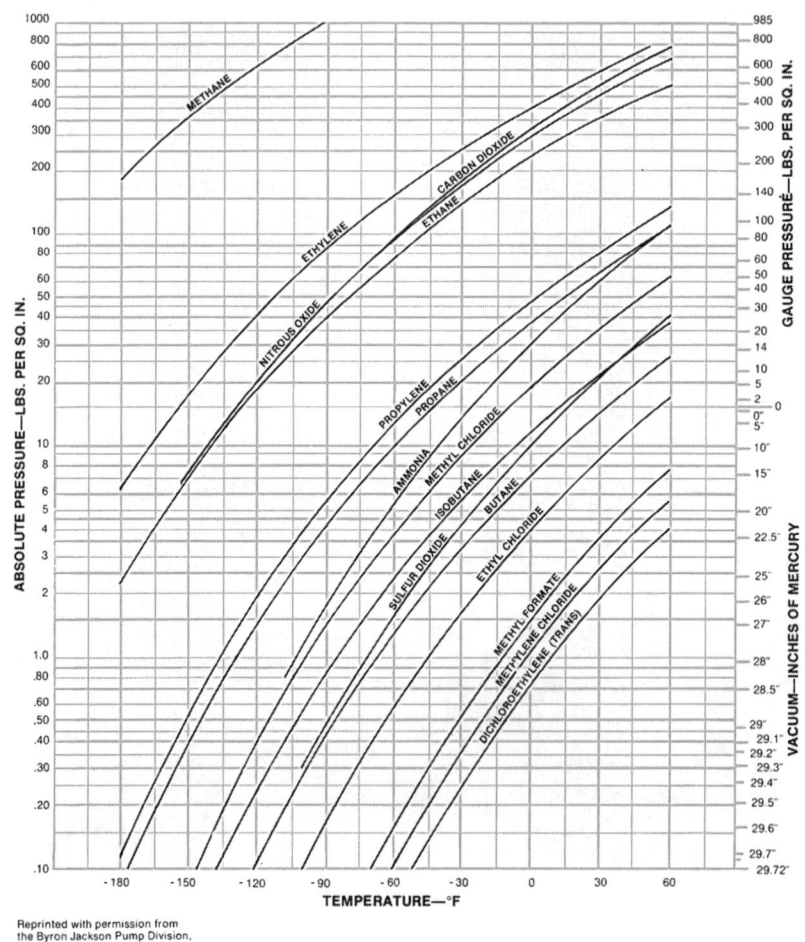

Figure 1.2 Vapor Pressure of Various Liquids, −180°F to 60°F *(Courtesy of the Hydraulic Institute)*

- Imperial gallons per minute = USgpm × 1.200
- Cubic meters per hour (m³/hr) = USgpm × 0.227
- Liters per second (L/sec) = USgpm × 0.063
- Barrels per day (one barrel = 42 gal) = USgpm × 34.290

The pump's flow capacity varies with impeller width, impeller diameter, and pump revolutions per minute (rpm).

Discharge Pressure

This is the pressure measured at the pump's discharge nozzle. Measurements may be stated in:

- Psig
- kg/cm^2
- Bars
- Kilopascals

Discharge Head

Measured in feet or meters, the discharge head is the same as the discharge pressure converted into the height of a liquid column.

Total Differential Head

The difference between the discharge head and the suction head is the total differential head (TDH), expressed in feet or meters.

Net Positive Suction Head

The net positive suction head (NPSH) available is the NPSH in feet available at the centerline of the pump inlet flange. The NPSH required (NPSHR) refers to the NPSH specified by a pump manufacturer for proper pump operation. (See Chapter 9.)

Density

This term refers to the mass per unit volume measured in pounds per cubic foot at 68°F or in grams per milliliter at 4°C.

Specific Gravity

Dividing the weight of a body by the weight of an equal volume of water at 68°F yields specific gravity (sp gr). If the data is in grams per milliliter, the specific gravity of a body of water is the same as its density at 4°C.

Suction Head

The height of a column of liquid upstream from the pump's suction nozzle's centerline is known as the suction head. It may also be the suction pressure, in psig, converted to suction head, in feet. Feet or meters measure suction head.

Table 1.2
Specific Gravity of Some Liquids

Liquid	Temperature °F	Specific Gravity	Weight (lb/gal)
Acetone	68.0	0.792	6.60
Aniline	68.0	1.022	8.51
Carbon tetrachloride	68.0	1.595	13.28
Coconut oil	59.0	0.926	7.71
Corn oil	59.0	0.925	7.70
Cottonseed oil	60.8	0.926	7.71
Ether	77.0	0.708	5.90
Fuel oil (No. 1)	60.0	0.800–0.850	6.70–7.10
Fuel oil (No. 2)	60.0	0.810–0.910	6.70–7.60
Gasoline	60.0	0.700–0.760	5.80–6.30
Glucose	77.0	1.544	12.86
Glycerin	32.0	1.260	10.49
Hydrochloric acid*	60.0	1.213	10.10
Kerosene	68.0	0.820	6.83
Linseed oil	68.0	0.930	7.80
Molasses	68.0	1.470	12.20
Olive oil	59.0	0.920	7.66
Soy bean oil	59.0	0.927	7.72
Sulfuric acid[†]	64.0	1.834	15.27
Tar	68.0	1.200	10.00
Seawater[††]	59.0	1.020	8.54
Water (0°C)	39.0	1.000	8.34
Water (20°C)	68.0	0.998	8.32

*43.4% solution
[†]87.0% solution
[††]May vary. Specific gravity of water in the Arabian Gulf is 1.03.

Suction Pressure

This refers to the pressure, in psig, at the suction nozzle's centerline. For instance, the pressure developed by a booster pump hooked up in series with a main pump is the suction pressure of the main pump measured at suction nozzle centerline.

Suction Lift

The maximum distance of a liquid level below the impeller eye that will not cause the pump to cavitate is known as suction lift. Because a liquid is not cohesive, it cannot be pulled. Instead, the pump impeller, pistons, plungers, or rotors form a partial vacuum in the pump. The atmospheric pressure (14.7 psi, or 34 ft) pushes the liquid into this partial vacuum. Because of mechanical losses in the pump, suction lifts are always less than 34 ft.

Velocity Head

This term refers to the kinetic energy of a moving liquid at a determined point in a pumping system. The expression for velocity head is in feet per second (ft/sec) or meters per second (m/sec). The mathematical expression is:

Velocity head (h_v) = $V^2/2g$

where:

V = liquid velocity in a pipe
g = gravity acceleration, influenced by both altitude and latitude. At sea level and 45° latitude, it is 32.17 ft/sec/sec.

Horsepower

The work a pump performs while moving a determined amount of liquid at a given pressure is horsepower (hp).

Cavitation

This implosion of vapor bubbles in a liquid inside a pump is caused by a rapid local pressure decrease occurring mostly close to or touching the pump casing or impeller. As the pressure reduction continues, these bubbles collapse or implode. Cavitation may produce noises that sound like pebbles rattling inside the pump casing and may also cause the pump to vibrate and to lose hydrodynamic efficiency. This effect contrasts boiling, which happens when heat builds up inside the pump.

Continued serious cavitation may destroy even the hardest surfaces. Avoiding cavitation is one of the most important pump design criteria. Cavitation limits the upper and lower pump sizes, as well as the pump's peripheral impeller speed.

Displacement

The capacity, or flow, of a pump is its displacement. This measurement, primarily used in connection with positive displacement pumps, is measured in units such as gallons, cubic inches, and liters.

Volumetric Efficiency

Divide a pump's actual capacity by the calculated displacement to get volumetric efficiency. The expression is primarily used in connection with positive displacement pumps.

Minimum Flow

The lowest continuous flow at which a manufacturer will guarantee a pump's performance is the pump's minimum flow.

Critical Speed

At this speed, a pump may vibrate enough to cause damage. Pump manufacturers try to design pumps with the first critical speed at least 20 percent higher or lower than rated speed. Second and third critical speeds usually don't apply in pump usage.

Minimum Flow Bypass

This pipe leads from the pump discharge piping back into the pump suction system. A pressure control, or flow control, valve opens this line when the pump discharge flow approaches the pump's minimum flow value. The purpose is to protect the pump from damage.

Area Classification

An area is classified according to potential hazards. For example, risks of explosions or fire may exist because of material processed or stored in the area.

Chapter 2

Pump Calculations

Friction

Various formulas calculate friction losses. Hazen-Williams wrote one of the most common for smooth steel pipe. Usually, you will not need to calculate the friction losses, because handbooks such as the *Hydraulic Institute Pipe Friction Manual* tabulated these long ago. This manual also shows velocities in different pipe diameters at varying flows, as well as the resistance coefficient (K) for valves and fittings.

To practice good engineering for centrifugal pump installations, try to keep velocities in the suction pipe to 3 ft/sec or less. Discharge velocities higher than 11 ft/sec may cause turbulent flow and/or erosion in the pump casing.

In the following problem, the following formula calculates head loss:

$H_f = K(V^2/2g)$ (Problem 2.1)

where:

H_f = friction head
K = friction coefficient
V = velocity in pipe
g = gravity (32.17 ft/sec/sec)

Find the total friction losses for a flow of 900 gpm of water at 68°F in a new 6-in. schedule 40 steel pipe, 250 ft long with two elbows, a check valve, and a gate valve. Valves and fittings are flanged. The elbows are 90°. Use the following *Hydraulic Institute Pipe Friction Manual* friction losses for pipe, valves, and fittings:

Equivalent Length (in ft)

Q = 900 gpm
V = 9.99 ft/sec
$V^2/2g$ = 1.55
  F_1 = pipe loss = 5.05 × 250/100   = 12.60
  F_2 = gate valve (K = 0.1) = 0.1 × 1.55   = 0.15
  F_3 = check valve (K = 2) = 2.0 × 1.55   = 3.10
  F_4 = elbows (K = 0.28) = 0.28 × 1.55 × 2 = 0.87
Total friction losses   = **16.72**

Head Calculations

In centrifugal pump calculations, the conversion of the discharge pressure to discharge head is the norm. Positive displacement pump calculations often leave given pressures in psi.

In the following formulas, W expresses the specific weight of liquid in pounds per cubic foot. For water at 68°F, W is 62.32 lb/ft^3. A water column 2.31 ft high exerts a pressure of 1 psi on 64°F water. Use the following formulas to convert discharge pressure in psig to head in feet:

• For centrifugal pumps

$$H \text{ (in ft)} = \frac{P \text{ (in psig)} \times 2.31}{\text{sp gr}}$$

• For positive displacement pumps

$$H \text{ (in ft)} = \frac{P \text{ (in psig)} \times 144}{W}$$

To convert head into pressure:

• For centrifugal pumps

$$P \text{ (in psi)} = \frac{H \text{ (in ft)} \times \text{sp gr}}{2.31}$$

• For positive displacement pumps

$$P \text{ (in psi)} = \frac{H \text{ (in ft)} \times W}{244}$$

The problem in the following example attempts to find the head of a salt water having a specific gravity of 1.03 at 68°F at 12 psig pressure, as well as a hydrocarbon with a specific gravity of 0.87 at the same temperature:

$$H = \frac{12 \times 2.31}{1.03} = 26.9 \text{ ft}$$

(Problem 2.2)

$$H = \frac{12 \times 2.31}{0.87} = 31.9 \text{ ft}$$

These two problems show that even though different liquids may display the same pressure, the head varies with the specific gravity of the liquids.

Problem 2.3 shows how to calculate the TDH of an end suction pump taking suction from a constant-level lake and discharging into an atmospheric tank (Figure 2.1). The *Hydraulic Institute Pipe Friction Manual* lists friction losses, $V^2/2g$ (velocity head conversion) values, and K values for valves and fittings.

Pump Calculations **11**

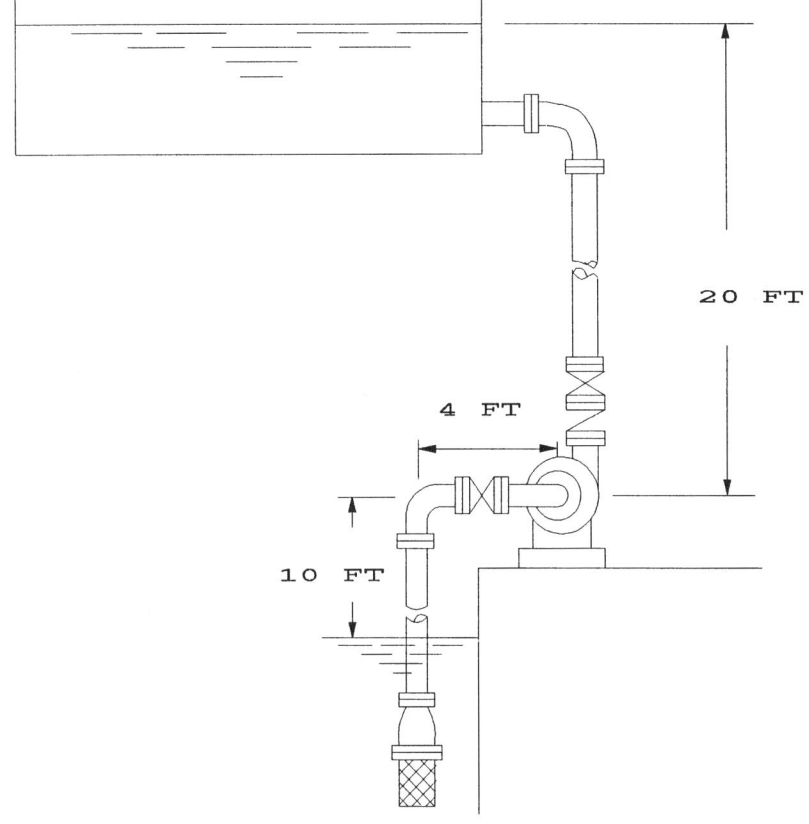

Figure 2.1 Suction Lift

In the following examples, valves and fittings are flanged. Elbows are standard 90°.

Given: (Problem 2.3)

Q = flow = 260 gpm
Liquid = water
  sp gr = specific gravity = 1.0
  t = temperature = 68°F
  ϕ_1 = diameter suction pipe = 6 in.
  ϕ_2 = diameter discharge pipe = 4 in.

12   *Practical Introduction to Pumping Technology*

	Equivalent Length	
S_1 = suction lift	= 10 ft	
V = velocity	= 2.89 ft/sec	
$V^2/2g$ = velocity head	= 0.13 ft	
E_1 = entrance (K = 0.50)	= 0.5 × 0.130	= 0.06 ft
F_1 = suction piping	= $\frac{14 \times 0.487}{100}$	= 0.07 ft
F_2 = foot valve (K = 0.8)	= 0.80 × 0.130	= 0.10 ft
F_3 = strainer (K = 0.8)	= 0.80 × 0.130	= 0.10 ft
F_4 = gate valve (K = 0.1)	= 0.10 × 0.130	= 0.01 ft
F_5 = 90° elbow (K = 0.28)	= 0.28 × 0.130	= <u>0.04</u> ft
		0.38 ft

Total suction lift = $S_1 + H_f$ = 10 + 0.38 = **10.38 ft**

Discharge Head	Equivalent Length	
S_1 = static head	= 20 ft	
V = velocity	= 6.55 ft/sec	
$V^2/2g$ = velocity head	= 0.67	
F_1 = discharge piping	= $\frac{11 \times 0.374}{100}$	= 0.04 ft
F_2 = check valve (K = 2.00)	= 2.00 × 0.667	= 0.13 ft
F_3 = gate valve (K = 0.15)	= 0.15 × 0.667	= 0.10 ft
F_4 = 2 elbows (K = 0.3)	= 0.30 × 0.667	= <u>0.20</u>
		0.47 ft

Total discharge head = $S_1 + P + H_f$ = 20 + 0.47 = **20.47 ft**

TDH = 20.47 + 10.38 = **30.85 ft**

Problem 2.4 shows how to calculate the TDH when a pump takes suction from an atmospheric tank and discharges into a pressurized manifold (Figure 2.2).

Given: (Problem 2.4)

Q = flow	= 320 gpm
Liquid = crude oil	
sp gr = specific gravity	= 0.92
t = temperature	= 68°F

	Equivalent Length
φ = pipe diameter	= 6 in.
S_1 = static head	= 4 ft

Pump Calculations 13

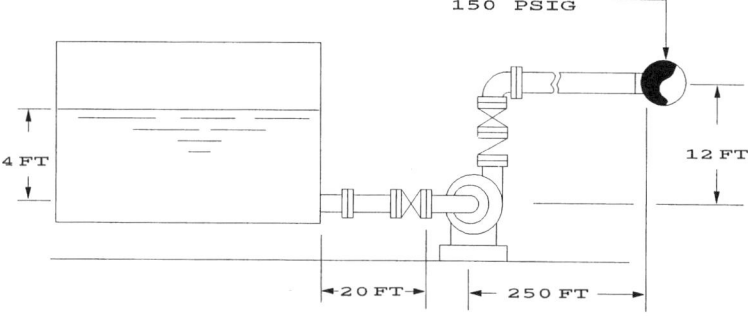

Figure 2.2 Flooded Suction, Atmospheric Tank

$$
\begin{aligned}
V &= \text{velocity} & &= 3.55 \text{ ft/sec} \\
V^2/2g &= \text{velocity head} & &= 0.2 \text{ ft} \\
E_1 &= \text{entrance (K = 0.5)} & &= 0.5 \times 0.2 & &= 0.10 \text{ ft} \\
F_1 &= \text{suction piping} & &= \frac{20 \times 0.719}{100} & &= 0.14 \text{ ft} \\
F_2 &= \text{gate valve (K = 0.1)} & &= 0.1 \times 0.2 & &= \underline{0.02} \text{ ft} \\
& & & & &0.26 \text{ ft}
\end{aligned}
$$

Total suction head = $S_1 - H_f$ = 4 – 0.26 = **3.74 ft**

Discharge Head Equivalent Length

$$
\begin{aligned}
\phi &= 4 \text{ in.} \\
V &= 8.06 \text{ ft/sec} \\
V^2/2g &= 1.01 \text{ ft} \\
P &= \text{manifold pressure} & &= 150 \text{ psig} \\
& \frac{150 \times 2.31}{0.92} & & 376.63 \text{ ft} \\
S_2 &= \text{static head} & &= 12 \text{ ft} \\
E_2 &= \text{exit (K = 0.5)} & &= 0.5 \times 1.01 & &= 0.50 \text{ ft} \\
F_1 &= \text{discharge piping} & &= \frac{250 \times 5.51}{100} & &= 13.77 \text{ ft} \\
F_2 &= \text{check valve (K = 2)} & &= 2 \times 1.01 & &= 2.02 \text{ ft} \\
F_3 &= \text{gate valve (K = 0.15)} & &= 0.15 \times 1.01 & &= \underline{0.15} \text{ ft} \\
& & & & &16.44 \text{ ft}
\end{aligned}
$$

Total discharge head = $P + S_2 + H_f$ = 376.63 + 12 + 16.44 = **405.07 ft**

TDH = 405.07 – 3.74 = **401.33 ft**

14 *Practical Introduction to Pumping Technology*

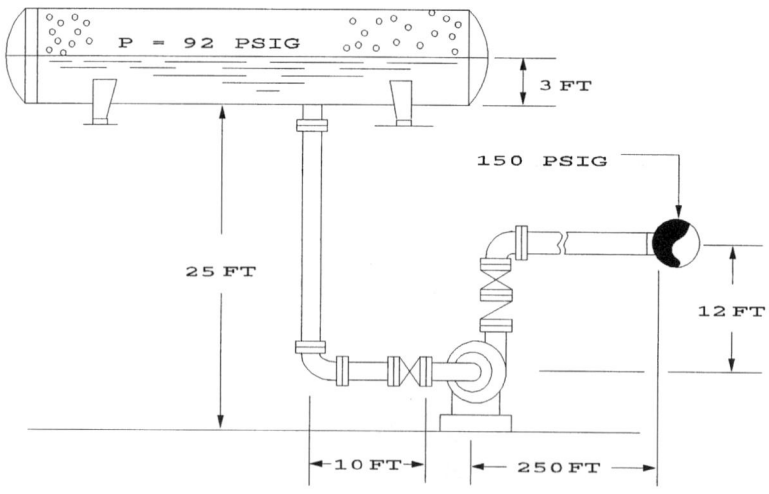

Figure 2.3 Suction From Pressure Vessel

Problem 2.5 shows how to calculate the TDH when a pump takes suction from a pressure vessel and discharges into a pressurized manifold (Figure 2.3).

Given: (Problem 2.5)

Q = flow	= 320 gpm
Liquid = crude oil	
sp gr = specific gravity	= 0.92
t = temperature	= 68°F

Suction	**Equivalent Length**	
ϕ = pipe diameter	= 6 in.	
P_1 = vessel pressure	= 92 psig	
$\dfrac{92 \times 2.31}{0.92}$	= 231 ft	
S_1 = suction head	= 25 ft	
V = velocity	= 3.55 ft/sec	
$V^2/2g$ = velocity head	= 0.2 ft	
E_1 = entrance (K = 0.5)	= 0.5 × 0.2	= 0.10 ft
F_1 = suction piping	= $\dfrac{35 \times 0.719}{100}$	= 0.09 ft
F_2 = gate valve (K = 0.1)	= 0.10 × 0.2	= 0.25 ft
F_3 = elbow (K = 0.28)	= 0.28 × 0.2	= 0.05 ft
		0.49 ft

Total suction head = $P_1 + S_1 - H_f$ = 231 + 25 − 0.49 = **255.51 ft**

Discharge Head	Equivalent Length

$\phi = 4$ in.
P_2 = manifold pressure = 150 psig
$\dfrac{150 \times 2.31}{0.92}$ = 376.63 ft
V = 8.06 ft/sec
$V^2/2g$ = 1.01 ft
S_2 = 12 ft
F_1 = discharge piping = $\dfrac{250 \times 5.51}{100}$ = 13.77 ft
F_2 = check valve (K = 2) = 2 × 1.01 = 2.02 ft
F_3 = gate valve (K = 0.15) = 0.15 × 1.01 = $\underline{0.15}$ ft
                                                             15.94 ft

Total discharge head = $P_2 + H_2 + H_f$ = 376.63 + 12 + 15.94 = **404.57 ft**

TDH = 404.57 − 255.51 = **149.06 ft**

Horsepower

While pushing a certain amount of liquid at a given pressure, the pump performs work. One horsepower equals 33,000 ft-lb/min. The two basic terms for horsepower are:

- Hydraulic horsepower
- Brake horsepower

One hydraulic horsepower equals the following:

- 550 ft-lb/sec
- 33,000 ft-lb/min
- 2,545 British thermal units per hour (Btu/hr)
- 0.746 kw
- 1.014 metric hp

To calculate the hydraulic horsepower (WHP) using flow in gpm and head in feet, use the following formula for centrifugal pumps:

$$\text{WHP} = \dfrac{\text{flow (in gpm)} \times \text{head (in ft)} \times \text{specific gravity}}{3{,}960}$$

When calculating horsepower for positive displacement pumps, common practice is to use psi for pressure. Then the hydraulic horsepower formula becomes:

$$\text{WHP} = \frac{\text{flow (in gpm)} \times \text{pressure (in psi)}}{1{,}714}$$

A pump's brake horsepower (BHP) equals its hydraulic horsepower divided by the pump's efficiency. Thus, the BHP formulas become:

$$\text{BHP} = \frac{\text{flow (in gpm)} \times \text{head (in ft)} \times \text{specific gravity}}{3{,}960 \times \text{efficiency}}$$

$$\text{BHP} = \frac{\text{flow (in gpm)} \times \text{pressure (in psig)}}{1{,}714 \times \text{efficiency}}$$

For Problem 2.6, calculate the BHP requirements for a pump handling salt water and having a flow of 500 gpm with 50 psi differential pressure. The specific gravity of salt water at 68°F equals 1.03. The pump efficiency is 85 percent. To use the first formula, convert the pressure differential to total differential head, TDH = 50 × 2.31/1.03 = 112 ft.

$$\text{BHP} = \frac{500 \times 112 \times 1.03}{3{,}960 \times 0.85} = 17.14\,\text{hp}$$

(Problem 2.6)

$$\text{BHP} = \frac{500 \times 50}{1{,}714 \times 0.85} = 17.16\,\text{hp}$$

Specific Speed

An impeller's specific speed (N_s) is its speed when pumping 1 gpm of liquid at a differential head of 1 ft. Use the following formula for specific speed, where H is at the best efficiency point:

$$N_s = \frac{\text{rpm} \times Q^{0.5}}{H^{0.75}}$$

where:

   rpm = revolutions per minute
   Q = flow (in gpm)
   H = head (in ft)

Pump specific speeds vary between pumps. No absolute rule sets the specific speed for different kinds of centrifugal pumps. Consider the following a rule of thumb for N_s ranges:

- Volute, diffuser, and vertical turbine = 500–5,000
- Mixed flow   = 5,000–10,000
- Propeller pumps   = 9,000–15,000

The higher the specific speed of a pump, the higher its efficiency. The best efficiency area covers a broad range. Pumps with low specific speeds have fairly flat head capacity curves (H-C curves). Pumps with low specific speed impellers commonly occur in pumps with small to average capacities and relatively high heads.

A low specific speed impeller has narrow channels between the vanes. Because most impellers are not precision cast, the danger of irregularities in these channels, which may cause pressure reduction and cavitation at low NPSHR values, always exists.

The H-C curve of high N_s pumps is steep, and the best efficiency range is narrow. Pumps with this type of impeller tend toward instability at low flows; these pumps require a high NPSH.

Suction Specific Speed

Also an impeller design characteristic, the suction specific speed (S) relates to the impeller's suction capacities. For practical purposes, S ranges from about 3,000 to 15,000. The limit for the use of suction specific speed impellers in water is approximately 11,000. Higher speeds demand unreasonably high NPSH, which if not met will cause cavitation around the impeller. The following equation expresses S:

$$S = \frac{\text{rpm} \times Q^{0.5}}{\text{NPSHR}^{0.75}}$$

where:

rpm = revolutions per minute
Q = flow in gpm
NPSHR = net positive suction head required

Affinity Formulas

The following formulas define relationships between impeller diameter, pump head, and brake horsepower. Becaue of inexact results, some deviations may occur in the calculations.

$$Q_2/Q_1 = D_2/D_1$$
$$H_2/H_1 = (D_2/D_1)^2$$
$$BHP_2/BHP_1 = (D_2/D_1)^3$$

where:

Q = flow
H_1 = head before change
H_2 = head after change
BHP = brake horsepower
D_1 = impeller diameter before change
D_2 = impeller diameter after change

The relation between speed (N) changes are as follows:

$$Q_2/Q_1 = N_2/N_1$$
$$H_2/H_1 = (N_2/N_1)^2$$
$$BHP_2/BHP_1 = (N_2/N_1)^3$$

where:

N_1 = initial rpm
N_2 = changed rpm

For Problem 2.7, change an 8-in. diameter impeller for a 9-in. diameter impeller, and find the new flow (Q), head (H), and brake horsepower (BHP) where the 8-in. diameter data are:

Q_1 = 320 gpm (Problem 2.7)
H_1 = 120 ft
BHP_1 = 12

The 9-in. impeller diameter data will be as follows:

$Q_2 = 320 \times 9/8$ = 360 gpm
$H_2 = 120 \times (9/8)^2$ = 152 ft
$BHP_2 = 12 \times (9/8)^3$ = 17

Chapter 3

Required Data for Specifying Pumps

Most pump buyers have fair ideas about what information vendors need to prepare quotations. For instance, a pump buyer might send out an inquiry giving the following data:

- Flow = 1,000 gpm
- Discharge pressure = 600 psig
- Suction pressure = 10 psig
- Liquid = water
- Specific gravity = 1.03

A vendor will need more information than the above to provide an adequate quotation. Most knowledgeable, honest vendors will request more data. The high specific gravity triggers an alarm to the buyer. Because of the high specific gravity, the vendor will probably assume the liquid is sea water, which requires certain construction materials. However, other solubles in the liquid may demand a different metallurgy. To assure you will receive a comprehensible quotation, give the vendor the following information:

- Pump capacity (gpm, L/sec, or m^3/hr)
- If pump will run in parallel, note whether the capacity given is for one pump only or for two or more pumps.
- Discharge pressure (psig, kg/cm^2, kilopascals, or bars)
- Suction pressure (psig, kg/cm^2, kilopascals, or bars)
- Liquid type
- Liquid characteristics
- Differential head (ft or m)
- If pump will be run at various capacities and head, buyer shall indicate so.
- NPSH available (ft or m)
- Liquid temperature (°F and °C)
- Maximum ambient temperature (°F and °C)
- Minimum ambient temperature (°F and °C)
- Vapor pressure (psia, kg/cm^2, kilopascals, or bars)
- Type of pump, eg., end suction, in-line, axially split, vertical turbine, submersible
- Material specifications

- Who will supply starter
- Who will supply the instruments required
- Proposed pump driver
- Proposed shaft sealing
- Area classification
- Base required, base plate, or oil field skid
- Pump type arrangement, whether fixed or portable
- Whether installed indoors or outdoors
- Who will mount the driver (driver vendor, pump vendor, or buyer)
- Who will supply the eventual control panel
- Who will supply eventual back-pressure valve and/or strainer (if pump is a vertical turbine pump)
- Who supplies the coupling

These data may be listed as above or as part of an attached data sheet. By not submitting all pertinent data with the inquiry, the buyer is at the mercy of the vendor. The buyer may get a pump that will give long, trouble-free service, but in all probability the buyer purchases trouble. Therefore, consider it extremely important that the engineer takes the time to write a comprehensive specification, however short, and prepares a data sheet.

The pump buyer must approach all purchases as if they were a new application. If the pump is a replacement, the tendency is to find the old data sheet and specification and to include it in the new purchase order. This can cause problems. Pumping conditions may have changed since the purchase of the last pump, and a review is always in order.

Chapter 4
Pump Types

A comprehensive way to arrange pumps into categories is to place them into three major groups, which may be divided into subcategories:

- Centrifugal pumps
- Positive displacement pumps

Centrifugal Pumps

In these pumps, the rotation of a series of vanes in an impeller creates pressure. The motion of the impeller forms a partial vacuum at the suction end of the impeller. Outside forces, such as the atmospheric pressure or the weight of a column of liquids, push fluid into the impeller eye and out to the periphery of the impeller. From there, the rotation of the high-speed impeller throws the liquid into the pump casing. Through the volute configuration of the pump casing or through diffuser vanes, the velocity head generated by the centrifugal motion of the impeller converts into a static head. The head a centrifugal pump generates depends on the velocity of the impeller, pump rpm, impeller diameter, and the number of impellers in series. A centrifugal pump will, in theory, develop the same head regardless of the fluid pumped. However, the pressure generated differs. The formula to convert feet to psi is:

$$\text{Liquid head (in ft)} = \frac{\text{psi} \times 2.31}{\text{sp gr}}$$

$$\text{Pressure (in psi)} = \frac{\text{head (in ft)} \times \text{sp gr}}{2.31}$$

With these formulas, one finds that a head of 10 ft of water with a specific gravity of 1.0 has a pressure of 4.33 psi and that the pressure of 10 ft of a hydrocarbon with a specific gravity of 0.85 equals 3.68 psi.

Three main categories of centrifugal pumps exist:

- Axial flow
- Mixed flow
- Radial flow

Any of these pumps can have one or several impellers, which may be:

- Open
- Closed
- Semi-open
- Single suction
- Double suction

Axial-Flow and Mixed-Flow Pumps

In axial-flow pumps, the pumped fluid flows along the pump drive shaft. The mixed-flow pumps give both an axial and a radial motion to the liquid pumped. These two types of high-volume, low-head pumps have steep H-C curves. Flow capacities may range from 3,000 gpm to more than 300,000 gpm. The discharge pressure seldom exceeds 50 psig. The pumps are used in low-head, large-capacity applications, such as:

- Municipal water supplies
- Irrigation
- Drainage and flood control
- Cooling water ponds
- Refinery and chemical plant offsite services

Radial-Flow Pumps

Most centrifugal pumps are of radial flow. These include:

- End suction pumps
- In-line pumps
- Vertical volute pumps (cantilever)
- Axially (horizontally) split pumps
- Multistaged centrifugal pumps
- Vertical turbine pumps

End Suction Pumps. The vast majority of centrifugal pumps are end suction pumps (Figures 4.1 & 4.2), also called overhung pumps. The name, end suction, stems from the fact that the suction flange is located at the eye, or the center, of the impeller. Discharge usually comes from the top of the pump, but on some end suction pumps the user may rotate the discharge nozzle to any position. The impeller attaches to the end of a horizontal shaft, supported by two radial bearings. These pumps are called overhung because the impeller is not between these two bearings, but at the end of the shaft.

To install the internals, the manufacturers split the pump casing into two major parts. The casing may be split either horizontally or vertically, the correct nomenclature being axial or radial split, respectively. An end suction pump has a radial split, with the casing of the volute type. End suction pumps seldom have more than one impeller. They are, in other words, single staged. End suction pumps have a large capacity range. The smallest pumps may only handle 5 gpm at a minimum head of 40 ft. The larger pumps may pump up to 60,000 gpm range at a head of more than 500 ft. The capacity of this type of pump is limited to what is practical to fabricate and to transport. Some of the larger end suction pumps are too big to be moved fully assembled and must be field erected.

Because the head generated by a centrifugal pump directly relates to the peripheral velocity of the impeller, the head generated is limited to what can be accomplished with one impeller. Most pump manufacturers limit the peripheral velocity to 300 ft/sec.

In-line Pumps. An in-line pump (Figure 4.3) has a vertical shaft. As the name implies, both the suction and the discharge nozzles sit on the same horizontal axis, or in line. The advantage of an in-line pump is that the piping configuration to and from the pump is simpler than for an end suction pump. Vertical electric motors drive most in-line pumps. The largest in-line pumps available hover around the 3,000 hp range. However, most in-line pumps are relatively small, and it is not common to see many above 200 hp.

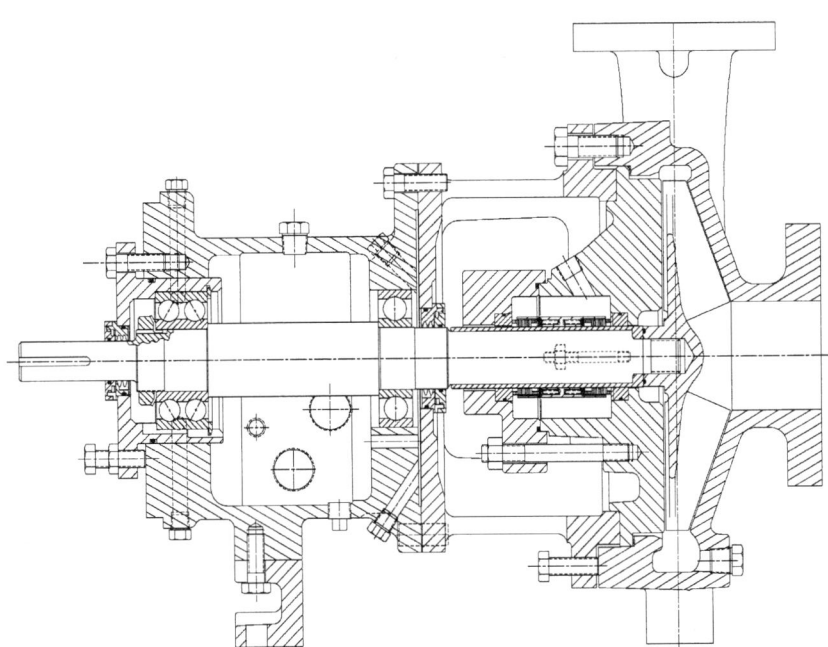

Figure 4.1 End Suction Pump (*Courtesy of Goulds Pumps Inc.*)

Figure 4.2 End Suction Pump (*Courtesy of Peerless Pump Co.*)

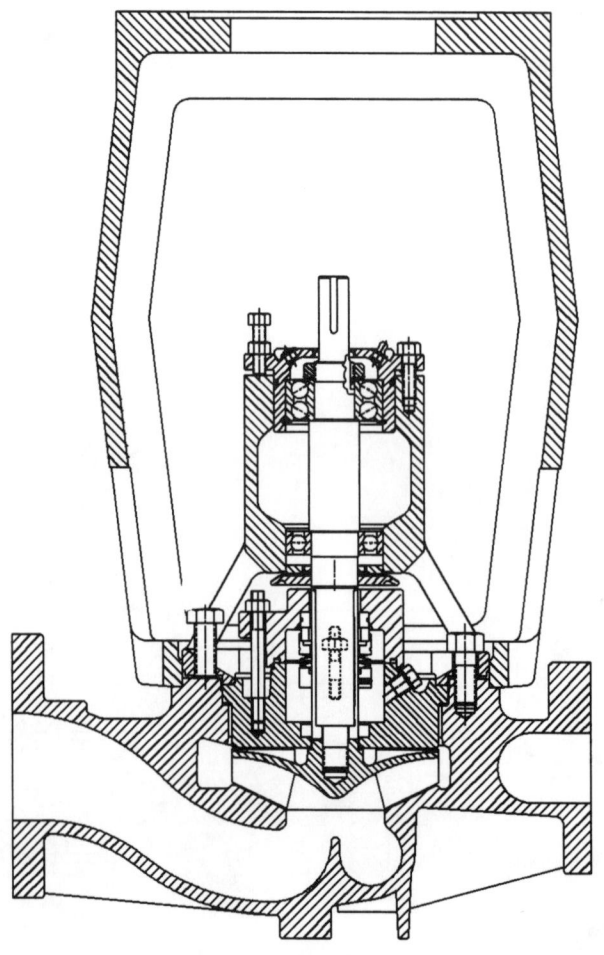

Figure 4.3 In-line Pump (*Courtesy of Goulds Pumps Inc.*)

Vertical Volute (Cantilever) Pumps. Also called cantilever pumps, vertical volute pumps (Figure 4.4) are basically of the same construction as horizontal centrifugal end suction pumps. The difference is the drive shaft assumes a vertical position. The entire pump submerges in the product. The driver, located above the liquid, connects to the impellers via a line shaft. Common uses for this type of pump include:

- Sewage pumps
- Sump pumps
- Chemical pumps
- Effluent pumps
- Non-clogging pumps

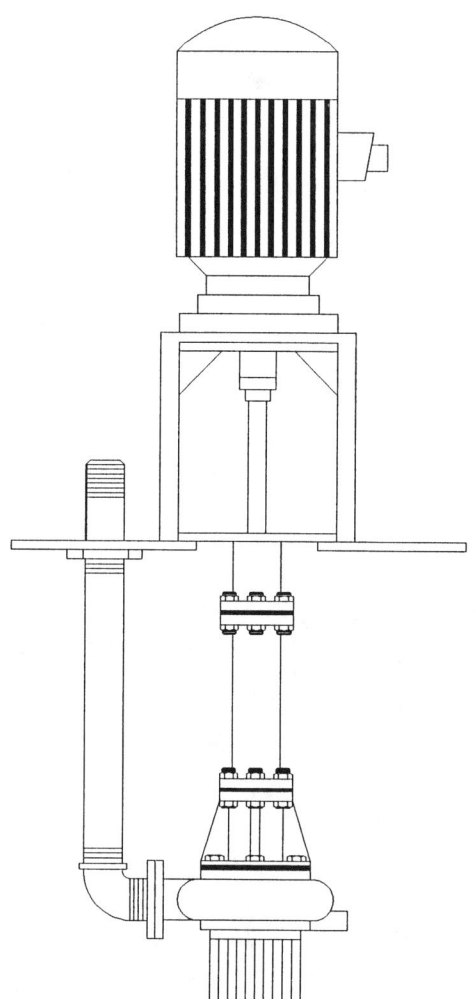

Figure 4.4 Vertical Volute (Cantilever) Pump

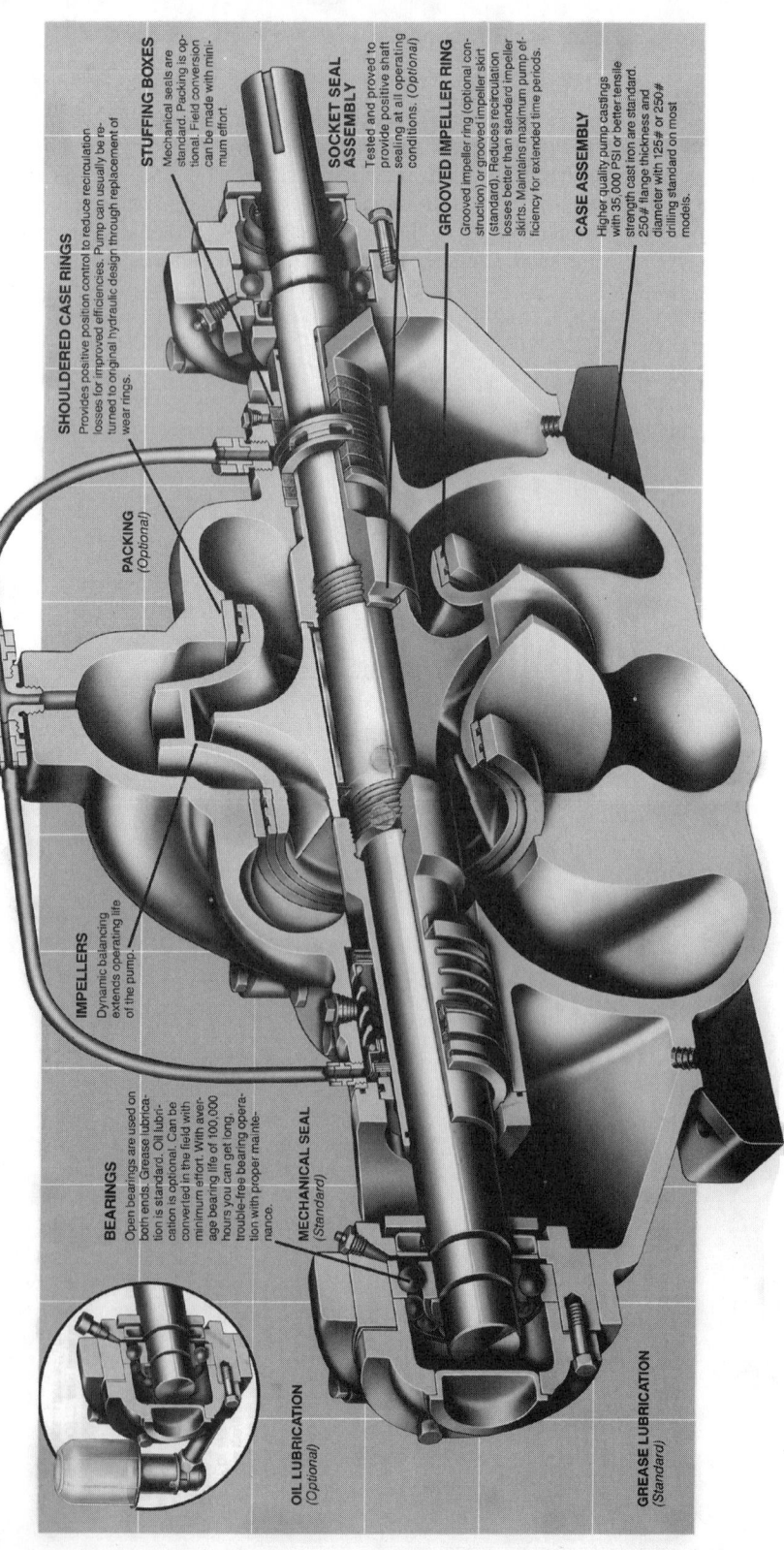

Figure 4.5 Axially Split Single-Stage Pump *(Courtesy of Peerless Pump Co.)*

Axially (Horizontally) Split Pumps. In an axially split pump (Figure 4.5), also called horizontal split-case or between bearing pump, the impeller lies between the two shaft bearings. The placement of the bearings make a sturdier construction. A pump with an axially split casing is also easy to repair and maintain. Lifting the upper part of the casing exposes the internals of the pump. Simply remove the entire rotating assembly to repair the pump, to balance it, and to trim the impellers. Manufacturers make axially split pumps in both single-stage and multi-stage versions, with either single or double suction impellers. Common uses for single-stage horizontal split-case pumps are:

- Water transport
- Process
- Pipeline
- Slurry
- Hydrocarbon handling
- Irrigation
- Fire protection
- Sugar processing
- Power generation plants

Multistaged Centrifugal Pumps. Multistaged pumps have two or more impellers placed in series. Three types of construction exist:

- Volute
- Diffuser
- Vertical turbine

Figure 4.6 shows a cross section of a volute multistaged pump (see Figure 4.7 also). To minimize the large thrust developed in these pumps, the impellers often sit opposite one another, as shown in the illustration. A thrust bearing absorbs the total thrust of the impellers. Because of construction restraints, the impellers in a multistaged diffuser pump are in series. Therefore a multistaged diffuser pump will always have a balancing device in addition to a thrust bearing. The thrust bearing carries the load during start-up and shutdown. The balancing devices carry most of the loads during operating speeds. The balancing device is either a balancing drum, a balancing disk, or a combination of both. This is one of the main disadvantages of a diffuser pump. If the balancing device fails, the thrust bearing is not large enough to absorb the horizontal thrust, and the pump will be severely damaged. Fortunately, temperature and vibration probes will detect incipient failure of the balancing device and trigger the pump into a shutdown mode.

The use of horizontal multistaged centrifugal pumps are common where conditions require a high discharge pressure combined with relatively low flow. Typical applications for this type of pump are:

- Boiler feed pumps
- Pipeline pumps

28   Practical Introduction to Pumping Technology

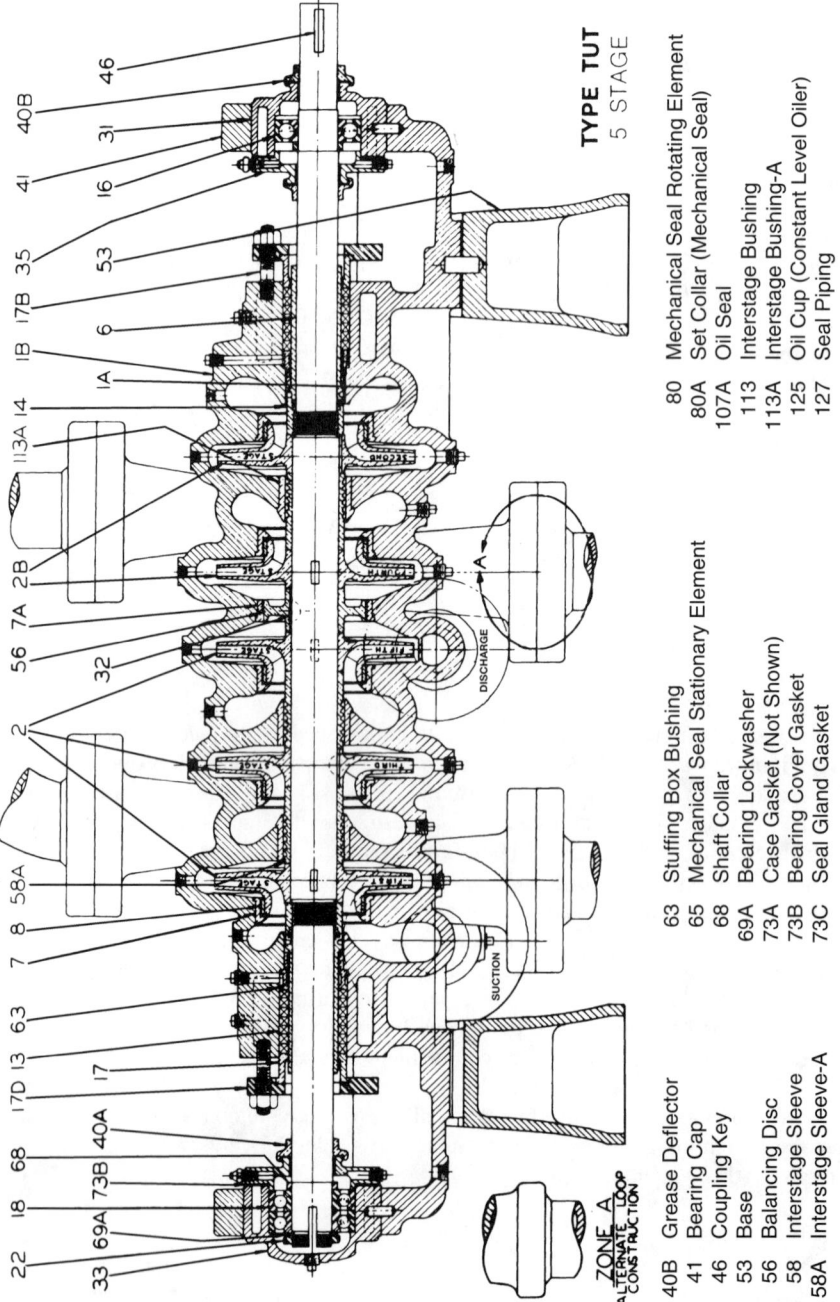

Figure 4.6 Axially Split Multistaged Volute Pump *(Courtesy of Peerless Pump Co.)*

- Chemical process pumps
- Mine dewatering pumps
- Reverse osmosis system charge pumps
- Oil field water injection pumps

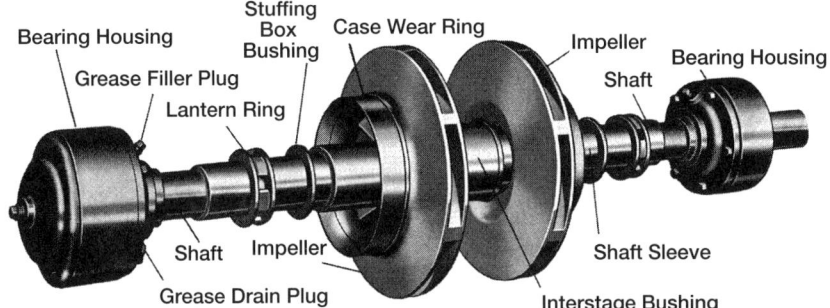

Figure 4.7 Axially Split Multistaged Volute Pump (*Courtesy of Peerless Pump Co.*)

Vertical Turbine Pumps. These pumps come either as line-shaft pumps or as submersible pumps. In both cases, the pump bowls submerge entirely in the liquid. The driver of a line-shaft pump (Figure 4.8), usually an electric induction motor or a diesel engine, is above the liquid. The impellers connect to the driver through a vertical shaft. Developed for water wells, people still widely use vertical turbine pumps for that purpose. They're also used as intake pumps by industries using large amounts of water, such as:

- Refineries
- Paper mills
- Power plants
- Chemical plants

Increasingly, vertical turbine pumps are replacing other pump types as cooling tower circulating pumps. Offshore oil and gas production facilities use this type of pump for fire water, utility, and process water intake pumps. The maximum setting depth for a line-shaft pump is about 400 ft.

A submersible pump (Figure 4.9) is a submerged vertical turbine pump with an electric motor attached to the bottom of the pump bowls. The whole assembly submerges in the liquid. An electric cable provides power from a source on the surface. Presently, you'll find this type of pump in applications such as:

- Deep oil and water wells
- Pipeline booster pumps
- Water intake pumps, both onshore and offshore

A vertical can pump (Figure 4.10) is a line-shaft vertical turbine pump enclosed in a casing or barrel. Use this type of pump configuration when not enough NPSH is available for horizontal centrifugal pumps moving volatile liquids and when the construction of a dry pit is not possible, as with hydrocarbons and other combustible liquids.

Positive Displacement Pumps

In this machine, the liquid flows into a contained space, such as a cylinder, plunger, or rotor. Then a moving piston forces the liquid out of the cylinder, increasing the pressure.

The use of positive displacement pumps is common in applications that require high discharge pressure and relatively low flow. The discharge pressure generated by a positive displacement pump is, in theory, infinite. If the pump is dead headed, the pressure generated will increase until either a pump part fails or the driver stalls from lack of power. The three basic types of positive displacement pumps are:

- Reciprocating pumps
- Rotary pumps
- Special-purpose pumps

Reciprocating Pumps

Four major types of reciprocating pumps exist:

- Steam pumps
- Power pumps
- Metering pumps
- Diaphragm pumps

Simplex pumps are reciprocating pumps with a single piston, plunger, or diaphragm. Names for multiple-cylinder pumps are:

- Duplex, with two cylinders
- Triplex, with three cylinders
- Quadruplex, with four cylinders
- Quintuplex, with five cylinders
- Multiplex, with many cylinders

Steam Pumps. Like all positive displacement pumps, the steam pump consists of a power end and a liquid end. The steam, or air end, may use steam or air as a power source. The liquid end consists of inlet and outlet ports, valves, and a piston or a plunger.

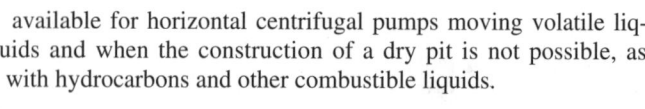

Figure 4.8 Vertical Turbine Line-Shaft Pump (*Courtesy of Peerless Pump Co.*)

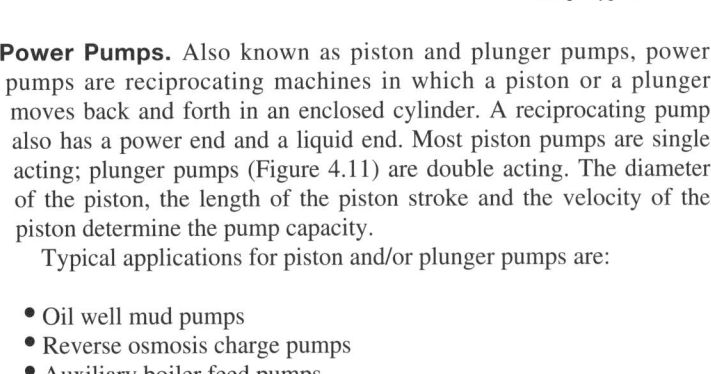

Power Pumps. Also known as piston and plunger pumps, power pumps are reciprocating machines in which a piston or a plunger moves back and forth in an enclosed cylinder. A reciprocating pump also has a power end and a liquid end. Most piston pumps are single acting; plunger pumps (Figure 4.11) are double acting. The diameter of the piston, the length of the piston stroke and the velocity of the piston determine the pump capacity.

Typical applications for piston and/or plunger pumps are:

- Oil well mud pumps
- Reverse osmosis charge pumps
- Auxiliary boiler feed pumps
- Pipeline pumps
- Oil field water injection pumps
- Slurry pumps
- Process pumps

Metering Pumps. Often chemicals, such as anticorrosion agents, flocculants, bacteriacide, or chlorine, are injected into a process stream. The two major types are:

- Electric-motor-driven metering piston pumps (Figure 4.12)
- Air- or gas-driven single or double diaphragm pumps.

Diaphragm Pumps. Several types of diaphragm pumps exist:

- Piston diaphragm pumps
- Double diaphragm pumps
- Gas/air-operated diaphragm pumps
- Gas/air-operated double diaphragm pumps

Like any positive displacement pump, the diaphragm pump also has a power end and a liquid end. The liquid end consists of a flexible membrane that pulsates in a shallow cylinder. The membrane functions like a reciprocating pump but with a much smaller stroke. The liquid only touches the diaphragm, the suction, and the discharge, which makes diaphragm pumps suitable for the following applications:

- Corrosive liquids
- Slurries
- Abrasive liquids

Figure 4.9 Submersible Vertical Turbine Pump (*Courtesy of Peerless Pump Co.*)

32   *Practical Introduction to Pumping Technology*

Figure 4.10 Vertical Can Pump (*Courtesy of Peerless Pump Co.*)

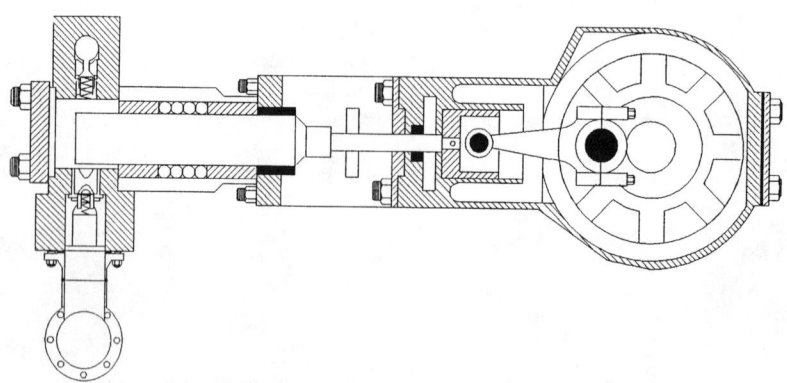

Figure 4.11 Triplex Plunger Pump

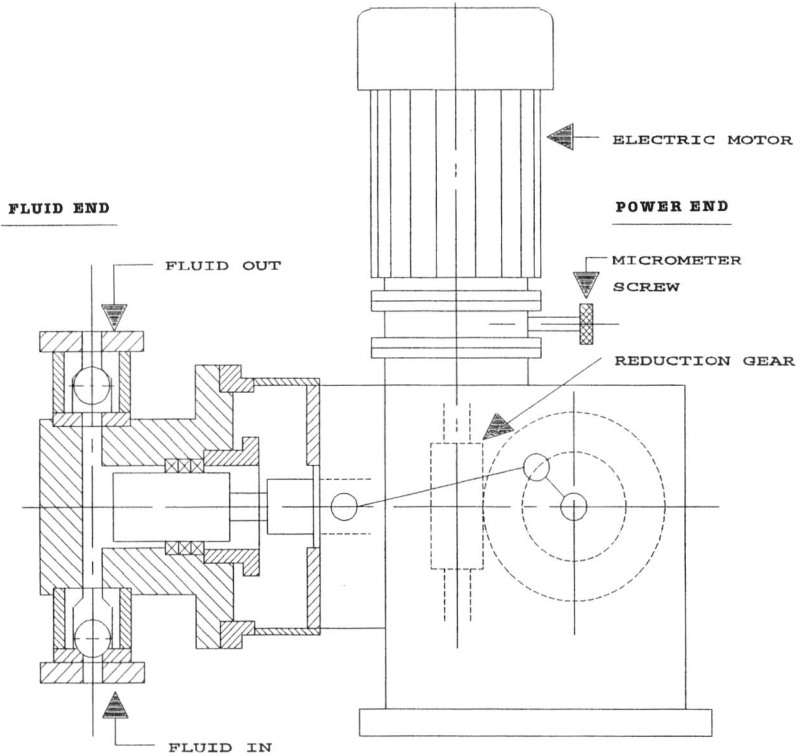

Figure 4.12 Electric-Motor-Driven Piston Metering Pump

- Food processing
- Chemical metering

Piston Diaphragm Pumps. In this type of pump (Figure 4.13), a crank attaches to the diaphragm. A hydraulic fluid moves the diaphragm up and down, which moves the piston in a reciprocating motion up and down. The piston is the liquid end of the pump. The movement of the piston regulates the amount of fluid pumped. Chemical metering pumps are often piston diaphragm pumps in which a micrometer screw regulates the length of the stroke of the piston.

Double Diaphragm Pumps. These function the same way piston diaphragm pumps do; double diaphragm pumps use two diaphragms for safety in case one of the membranes ruptures. Pumps handling highly flammable or toxic liquids often have two diaphragms.

Gas/Air-Operated Diaphragm Pumps. In a gas- or air-operated diaphragm pump, gas or air valves move the diaphragm up and down. A solenoid valve regulates the

34 *Practical Introduction to Pumping Technology*

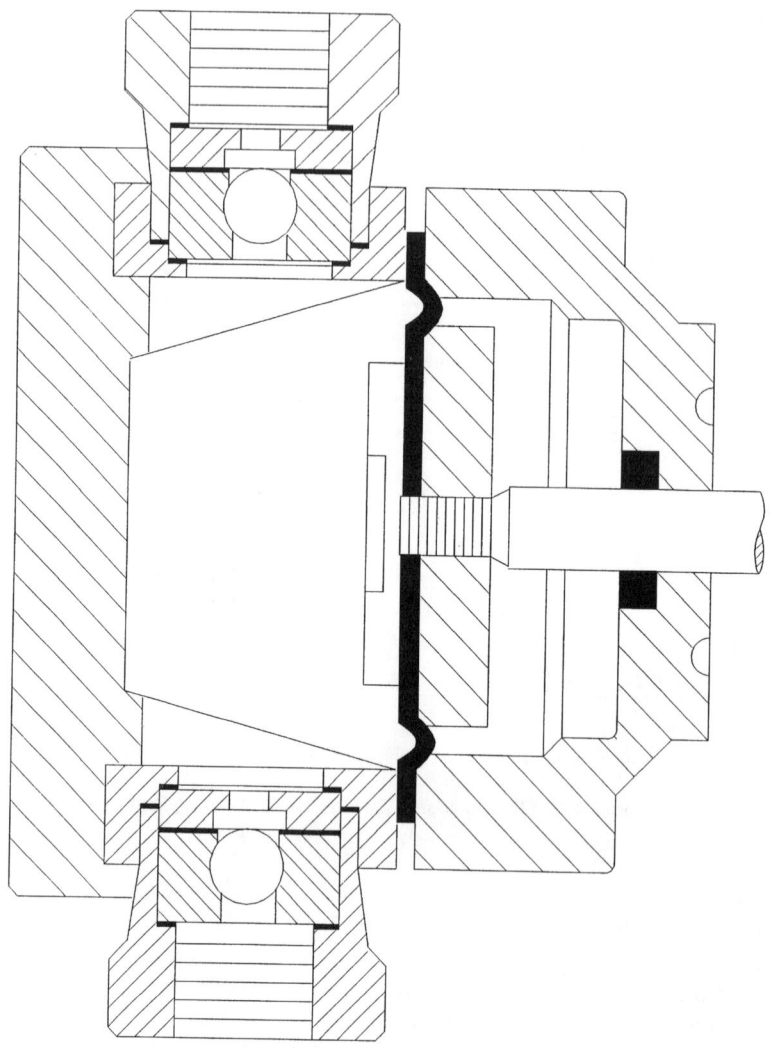

Figure 4.13 Single Diaphragm Pump

movement of the diaphragm. In another design, a double-acting air motor, instead of a mechanically operated crankshaft, drives the hydraulic fluid.

Gas/Air-Operated Double Diaphragm Pumps. This type of pump has two diaphragms joined together by a connecting rod. Air or gas pressure applied to the back of one diaphragm forces the product out of the liquid chamber into the discharge manifold. As the two diaphragms are connected, the other diaphragm is pulled toward the center of the pump. This action causes the other side to draw product into the pump on a suction stroke. At the end of the stroke, the air mechanism automatically shifts the air pressure to reverse the action of the pump. Ball valves open and close automatically to fill and empty chambers and to block backflow.

Reciprocating pumps may experience vapor lock when insufficient NPSH is available. The flow in reciprocating pumps pulsates, and therefore most pump applications require pulsation dampeners on both the discharge and suction sides of the pumps.

Rotary Pumps

This positive displacement machine has a rotary displacement element, such as gears, screws, vanes, or lobes. Each compartment between the dividing elements will hold a determined volume of fluid. As the first compartment fills with liquid, the fluid in the last compartment flows into the discharge piping. The pump capacity depends on the size of the compartments and the rotational speed of the pump. Typical rotary pumps include:

- Gear pumps
- Vane pumps
- Screw pumps
- Progressive cavity pumps
- Lobe pumps

External Gear Pumps. The fluid end of an external gear pump (Figure 4.14) consists of two herringbone gears of equal diameter mounted on a drive shaft and an idler shaft. The product flows into the suction end. It then moves through the intermeshing gears to the discharge end, where it discharges under higher pressure. The volume of product depends on the size of the gears and the speed of the rotating assembly. External gear pumps may move from 10 gpm up to more than 2,000 gpm. Gear pumps generally need liquids with some viscosity, such as hydrocarbon and food products.

Internal Gear Pumps. A small gear, mounted eccentrically, drives a larger one in an internal gear pump (Figure 4.15). When the gears rotate, they produce pockets into which the product moves at higher and higher pressures until forced out at the discharge end. An internal gear pump costs more money than an external one but can handle more viscous fluids. Pressures and flows are comparable between the two.

Sliding Vane Pumps. A slotted rotor mounted in a circular casing is the basic configuration of a sliding vane pump (Figure 4.17). The centrifugal force of the rotor causes the stiff vanes to slide in and out of the slots in the rotor. The vanes glide across the casing, forming a seal. Product flows into the pumps through the largest space between the vanes. The volumes in the adjacent spaces are progressively smaller. The discharge end appears where the volume in the spaces is smallest. The sliding vane pump suits both viscous and nonviscous fluids. Sliding vane pumps cannot handle dirty or gritty liquids. Vacuum ser-

36 *Practical Introduction to Pumping Technology*

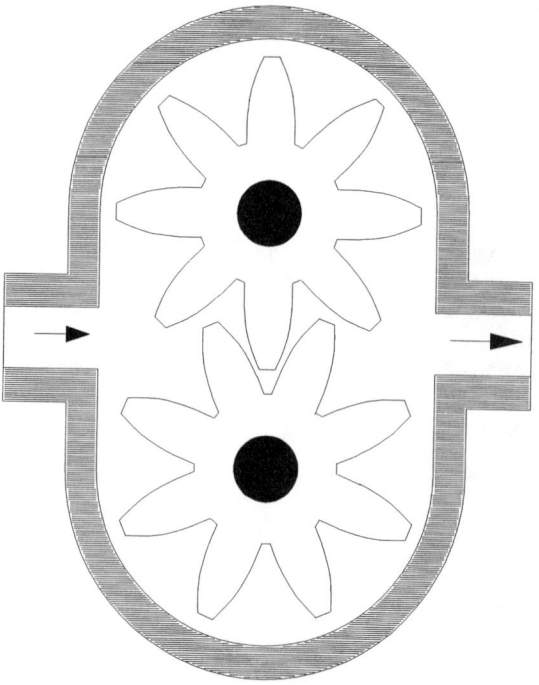

Figure 4.14 External Gear Pump

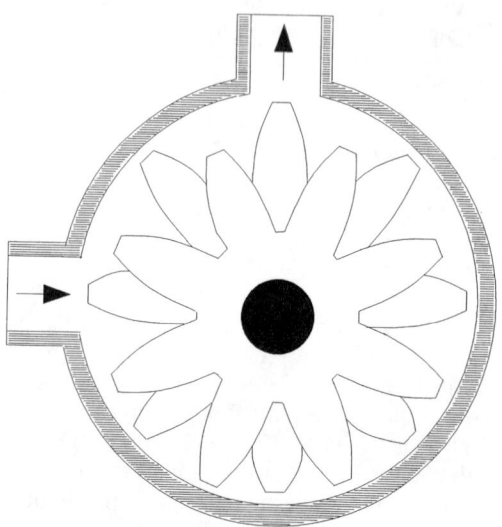

Figure 4.15 Internal Gear Pump

Pump Types 37

Figure 4.16 Seal-Less Rotary Gear Pump (*Courtesy of Roper Pump Co.*)

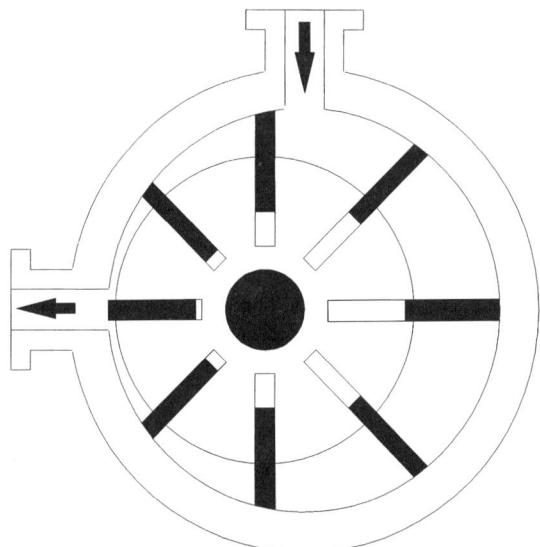

Figure 4.17 Sliding Vane Pump

vices provides another use for sliding vane pumps. A variation of this design is the flexible vane pump, which uses flexible vanes instead of rigid ones.

Twin Screw Pumps. This pump (Figure 4.18) consists of a driver screw and an idler screw run in a liner with precision tolerances. The rotation of the screws creates a continuous series of sealed chambers, moving the fluid axially from suction to discharge. The constant, practically pulse-free movement eliminates the need for pulsation dampeners. The highest discharge pressures achieved are around

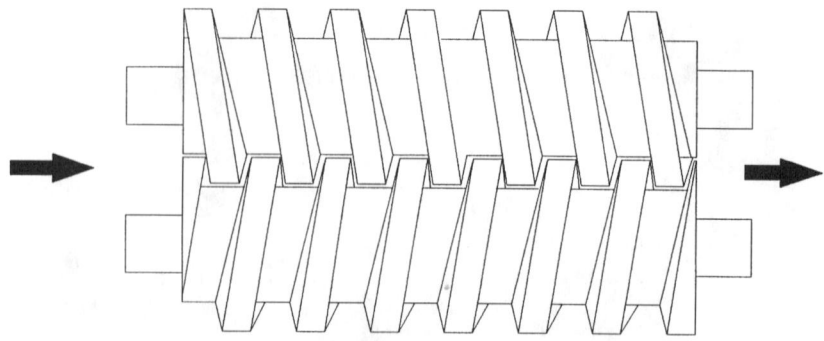

Figure 4.18 Twin Screw Pump

300 psig. Capacities can reach 2,000 gpm. Maximum acceptable viscosity is around 1,500 centistoke. Screw pumps with three screws are also available for conditions requiring higher capacities. Flows as high as 6,000 gpm are possible, with discharge pressures around 200 psig.

Progressive Cavity Pumps. These pumps handle a wide range of fluids, from clear water to slurries. The essential component of a progressive cavity pump (Figure 4.19) is a single helix rotor turning eccentrically within a double helix stator of twice the pitch. As the rotor turns, cavities form, which progress toward the discharge end of the pump. This action provides a pulse-free flow. Maximum pump capacity is around 1,000 gpm, with discharge pressures up to 300 psig. Progressive cavity pumps prime themselves.

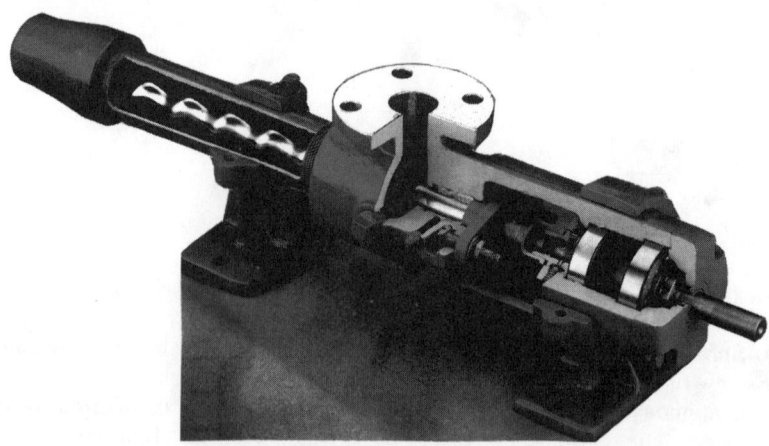

Figure 4.19 Progressive Cavity Pump (*Courtesy of Roper Pump Co.*)

Lobe Pumps. The food and beverage processing industry favors the use of lobe rotor pumps, partially because these pumps can handle both low and high viscosity fluids. Capacities range from 20 gpm to more than 1,000 gpm. This type of pump functions similarly to gear pumps. A rotor assembly consisting of two rotors, each mounted on its own shaft, provides the pumping action. The rotors may have single or multiple lobes (Figure 4.20). The rotor assembly housing consists of one chamber. The shape of the lobes does not permit one lobe to drive the other. Instead, timing gears control the movement of the rotors.

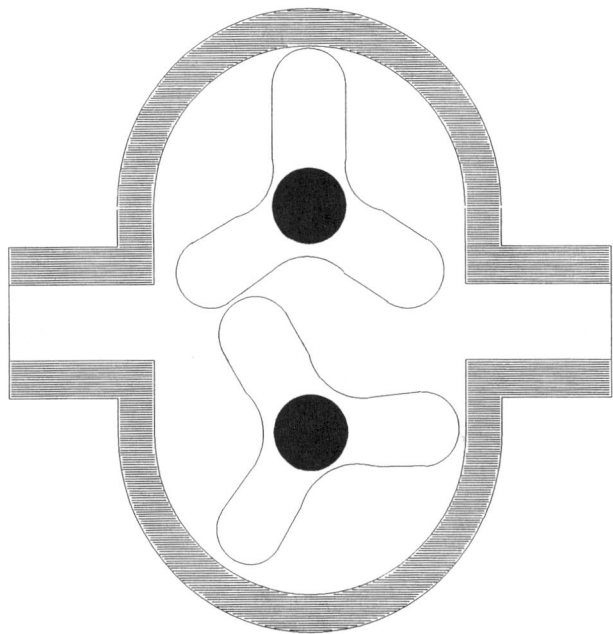

Figure 4.20 Multiple Lobe Pump

Special-Purpose Pumps

Several pumps defy qualification as either centrifugal or positive displacement pumps. This category comprises rarer pumps, but it doesn't hurt to be familiar with some of them, for instance:

- Archimedes screw pumps
- Pitot tube pumps
- Peristaltic pumps

Archimedes Screw Pumps. In a typical Archimedes screw pump (Figure 4.21), a helical screw rotates in a stationary trough. This is the oldest pump still in use. The ancient Egyptians used it to lift water from the Nile to irrigate surrounding fields. More recently, this pump is enjoying a renaissance. This specialty pump is used in applications requiring large flows and low heads. In its original form, the inefficient pump had a lot of backflow. In a modern improvement, a cylinder contains conically pitched helical flights, welded to the cylinder's walls. The whole assembly rotates. No loss of efficiency due to leakage or backsplashing occurs. A typical use is as a charge pump for tilted-pad separators, a tank where oil vestiges separate from water. There, centrifugal pumps have a distinct disadvantage because the high revolutions of these pumps tend to emulsify the oil, making it difficult to separate the oil from the water. Flows as high as 10,000 gpm are possible. Other applications for the Archimedes screw pumps include:

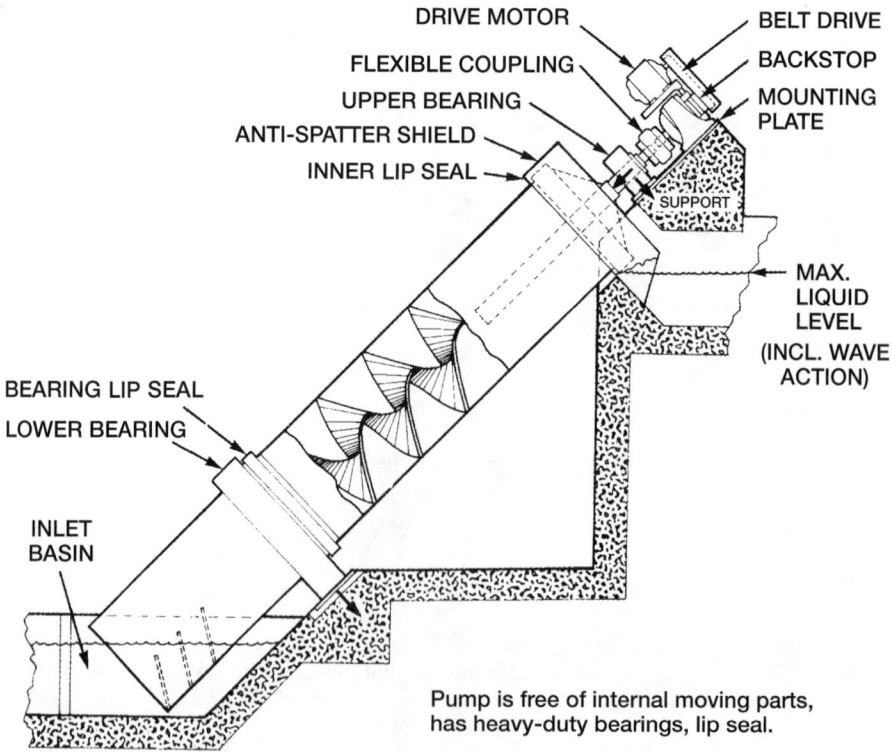

Figure 4.21 Archimedes Screw Pump (*Courtesy of U.S. Filters/CPC*)

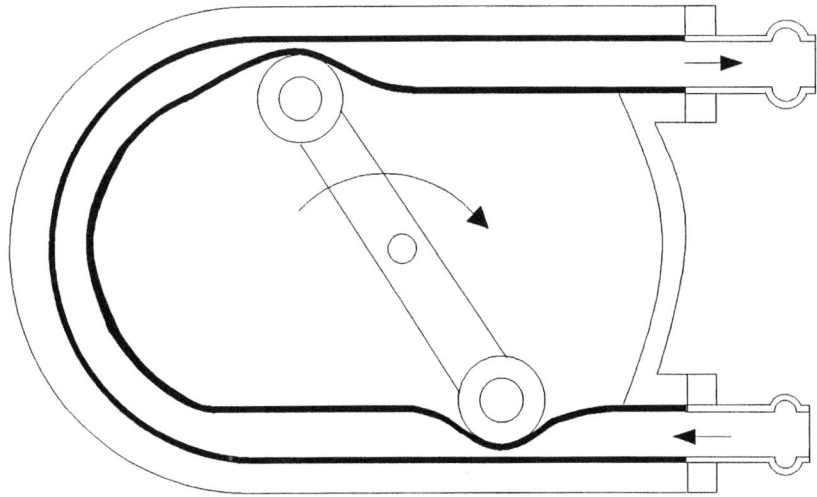

Figure 4.22 Peristaltic Pump

- Raw sewage lift stations
- Returning activated sludge
- Effluent lift stations
- Storm water lift stations
- Industrial waste pump stations

Pitot Tube Pumps. In this pump without an impeller, the liquid flows through a pitot tube into a casing, which rotates at a high speed. These pumps deftly suit applications with low viscosity liquids requiring low flows and high heads. Capacity limitations hover around 500 gpm. The pumps can generate heads up to 3,000 psig. Because the pumps often spin at more than 10,000 rpm, they have high NPSH requirements.

Peristaltic Pumps. A rolling action of a cam that squeezes the liquid through a soft plastic or rubber tube generates the flow in this pump (Figure 4.22). Transporting slurries embodies the most common use of this pump.

Chapter 5
Specifications

The buyer should always include a data sheet and/or a specification with an inquiry; even just one page will suffice. Depending on how costly and/or complicated the pump package will be, the buyer may also prepare project specifications for the following:

- Pump driver, which may be an electric motor, an internal combustion engine, a gas turbine, or a steam turbine
- Instruments for packaged equipment
- Structural steel skids
- Gears
- Couplings
- Equipment vibration
- Noise limits

The buyer must analyze the need to include any of these specifications. Obviously, a specification is not necessary for a small 1 hp water pump. A simple one-page data sheet is plenty. On the other hand, a large 6,000 hp boiler feed pump may require all the above-mentioned specifications.

Data Sheets

The buyer needs to fill out a data sheet before writing the project specification. This data sheet is the most important document in the inquiry; it will follow the pump through its life. Some companies design their own data sheets. Others use copies of the pump data sheets published in *American Petroleum Institute Standard 610, Centrifugal Pumps for General Refinery Services (API Standard 610)*, or variations thereof. For ANSI (American National Standards Institute) and ANSI/AWWA (American Water Works Association) pumps, much simpler data sheets of about one sheet are perfectly adequate.

The upper right corner of the data sheet provides a place for registering revisions. A common practice is to use letters for revisions before issuing the purchase order.

Data requiring completion by the vendor may have an asterisk (*) in front of it. When the data sheet comes back to the buyer with the vendor's quote, the vendor has completed the data sheet as per instructions. The buyer may not accept some of the vendor's features. He or she will discuss eventual changes with the vendor in a preaward meeting and finally issue a corrected data sheet with the purchase order.

This revision accompanying the purchase order will have a number, usually 0. All subsequent revisions will also have numbers. For instance, a numbering sequence as follows represents the standard throughout the industry:

- Revision A—Issue for approval
- Revision B—Issue for bid
- Revision 0—Issue for purchase
- Revision 1—Changed impeller diameter from 3.75 in. to 3.68 in.

Revision 0 is by no means the last revision. Every time Operations changes some pump feature, the change will appear on the data sheet as a revision. In this case, Revision 1 denotes trimming the impeller from 3.75 in. to 3.68 in.

Specifications

A project's specifications shall be in written form and will refer to the latest revisions of applicable industry standards, such as:

- *API Standard 610, Centrifugal Pumps for General Refinery Applications.*
- *API Standard 613, Special-Purpose Gear Units for Refinery Services.*
- *API Standard 614, Lubrication, Shaft Sealing, and Control Oil Systems for Special-Purpose Applications.*
- *API Standard 616, Combustion Gas Turbines for General Refinery Services.*
- *API Standard 671, Special-Purpose Couplings for Refinery Services.*
- *API Standard 674, Positive Displacement Pumps—Reciprocating.*
- *API Standard 675, Positive Displacement Pumps—Controlled Volume.*
- *API Standard 676, Positive Displacement Pumps—Rotary.*
- *ANSI-B73.1, Specification for Horizontal Centrifugal Pumps for Chemical Process.*
- *ANSI-B73.2, Specification for In-line Centrifugal Pumps for Chemical Process.*
- *ANSI/AWWA E101, Standard for Vertical Turbine Pumps, Line-Shaft, and Submersible Types.*
- *Hydraulic Institute Standards for Centrifugal, Rotary, and Reciprocating Pumps.*

The most common, and maybe the only correct way to make a project specification, is to write it around the applicable industry specification. For instance, *API Standard 610* contains references for mechanical seals. In this case, the buyer first fills out either an *API Standard 610* data sheet or one of his or her company's standard data sheets. The buyer writes the project specification, as shown in Appendix 1, with numbered subheadings. The buyer then adds the corresponding *API Standard 610* in parentheses next to the applicable project specification number. Then he or she goes through the latest edition of *API Standard 610*, noting the paragraphs with a bullet next to them (•). The bullets mark the information the buyer should supply.

For instance, paragraph 2.7.1.2 of *API Standard 610* says:

> Mechanical seals shall be of the single-balanced type (one rotating face per seal chamber) with either a sliding gasket or a bellows between the axially

moveable face and the shaft sleeve or housing. Unbalanced seals shall be furnished when specified or approved by the purchaser, or they may be recommended by the vendor if required for the service. Double seals have two rotating faces per chamber, sealing in opposite directions, and tandem seals have two rotating faces per chamber, sealing in the same direction.

If the vendor does in fact want double mechanical shaft seals with tungsten carbide against silicon carbide faces, it shall add that to its specification, in sequence with the correct paragraph number and subsection as follows:

5.3 (2.7.1.2) (Clarification)

Mechanical seals shall be double, balanced type with two rotating faces per box, facing in opposite directions. Seal faces shall be tungsten carbide against silicon carbon. Seals shall be API Code BDPFX.

Additions to industry specifications may be shown as completely new paragraphs, or as additions or subtractions of existing ones, such as:

2.8   NPSH (4.3.4.1) (Addition)
2.8.2 Vendor shall guarantee 40,000 hr impeller life.

The addition may be more detailed and may be expressed as follows:

2.8   NPSH (4.3.4.1) (Addition)
2.8.2 NPSHR data shall be taken at the following four points: minimum continuous flow, midway between minimum and rated flow, rated flow, 110 percent of rated flow. The NPSHR curve shall be for 40,000 hr impeller life.

Both paragraphs 2.8.2 say the same thing. In the first example, the writer specifies only the requirements for a guaranteed 40,000 hr impeller life. In the next example, the writer repeats the content of *API Standard 610*, paragraph 4.3.4.1, and then adds a request for 40,000 hr impeller life. Both methods are correct.

When the buyer regularly buys pumps not covered by industry standards but still needs a specification, he or she may prepare a simple company pump specification. With each purchase, the buyer can then modify this specification according to his or her needs. Appendix 1 shows such a typical company specification for centrifugal pumps.

Chapter 6

Pump Curves

Centrifugal Pump Curves

Head Capacity Curves

After a manufacturer designs and builds a pump, it checks the performance by testing the pump at various flows. Because centrifugal pumps are by no means high-technology machines, the results may differ from the design parameters. The results are plotted on a curve on which the y-axis represents the pump head and the x-axis the pump flow, or capacity. This curve is called the head capacity, or H-C, curve. To this curve the manufacturer adds more information, also in the shape of curves, including:

- The efficiency curves, a series of curves that show the internal losses of the pump at different capacities. The curves are shown as percent efficiency. The highest efficiency is the best efficiency point (BEP).
- The BHP curve, which shows the brake horsepower that the pump expands at different flows, from 0 gpm to maximum flow.
- The NPSH curve, which shows the NPSHR for the pump to function properly at a 2 percent head drop.

The pump manufacturer usually shows at least three different head capacity curves on the same sheet for each model. The different curves show different-sized impellers that fit within that particular pump case. Figure 6.1 shows a typical head capacity curve with the information normally included. In multistaged centrifugal pumps, the manufacturer usually shows only the first-stage pump curve in its general literature. If the client requests a certified head capacity curve, the manufacturer will draw a curve representing the sum of all the stages.

The characteristics of a centrifugal pump provide that, at constant speed (rpm) and with a specific impeller diameter, the curve will not change, regardless of the properties, weight, and type of liquid pumped. However, the curve will change if either the speed or the impeller diameter changes.

Problem 6.1 shows how the performance curve shown in Figure 6.2 will change if the pump rpm changes from 3,560 to 4,200 through a speed-increasing gear.

Q_1 = 1,375 gpm
H_1 = 110 ft (Problem 6.1)
Q_2 = 1,375 × 4,200/3,560 = 1,622 gpm
H_2 = 110 × (4,200/3,560)^2 = 153 ft

46 *Practical Introduction to Pumping Technology*

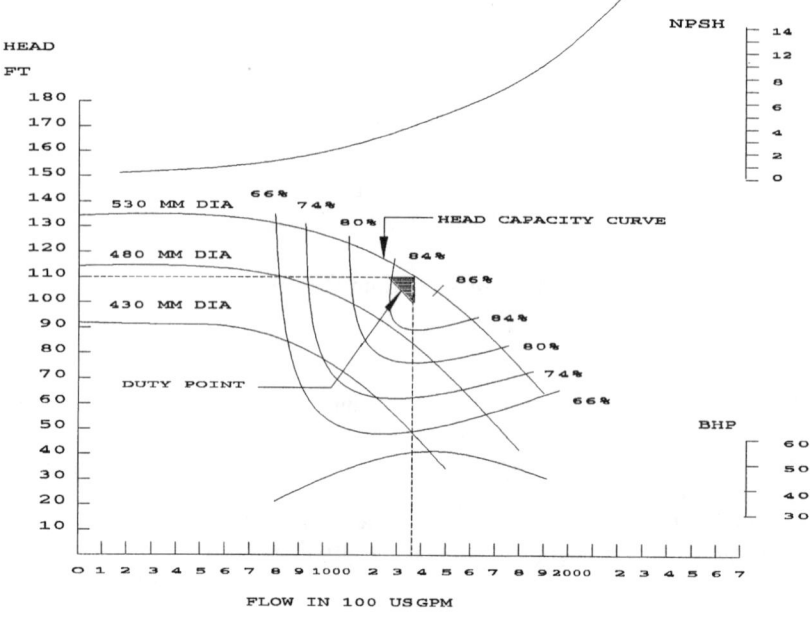

Figure 6.1 Head Capacity Curve

When selecting a pump, the buyer specifies the differential head and the capacity. This is the operating point of the pump. Figure 6.2 shows a curve drawn for a specified application.

In this case, the buyer has requested a pump that will pump freshwater with a specific gravity of 1.0 and with a capacity of 1,375 gpm at a differential head of 110 ft.

Often the pump vendor does not draw an individual curve for each pump request received. Instead, the vendor marks the operating point of the existing pump curves for the particular model the vendor offers for this application. If the operating point does not fit onto one of the existing impeller diameters, the vendor adds a dotted curve that represents the trimmed impeller it proposes to use. Figure 6.3 shows how the BHP will change with the change in the impeller diameter.

Other factors will change the characteristics of the performance curves for a given impeller diameter. One is the viscosity of the liquid. The National Hydraulic Institute publishes viscosity correction charts for centrifugal pumps. Viscosities over 250 Seconds Saybolt Universal (SSU) need viscosity corrections because capacity, head, and efficiency decrease while the BHP increases.

Variations in rpm (Figure 6.4) also change the pump curve according to the relationships discussed in Chapter 2, where Q is the flow, H the head, and N the speed:

$Q_2/Q_1 = N_2/N_1$

$H_2/H_1 = (N_2/N_1)^2$

Pump Curves **47**

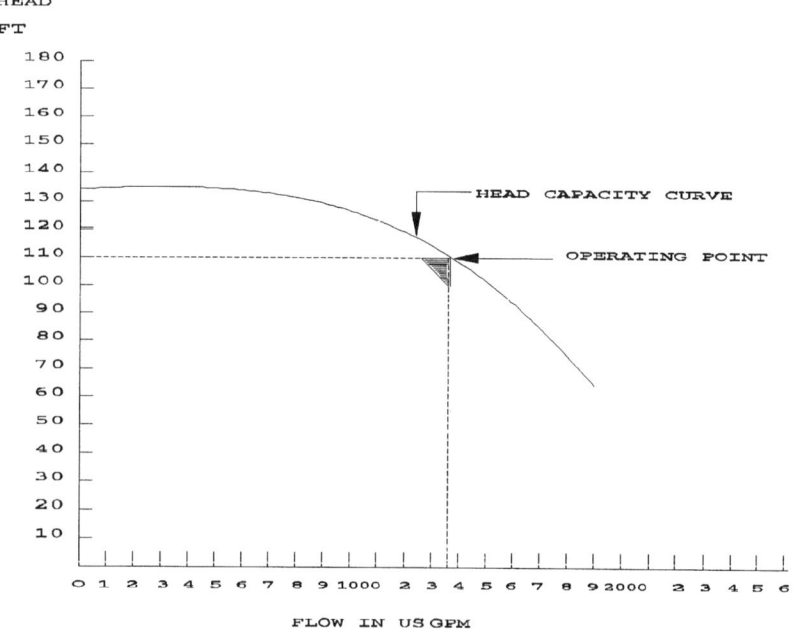

Figure 6.2 Individual Head Capacity Curve

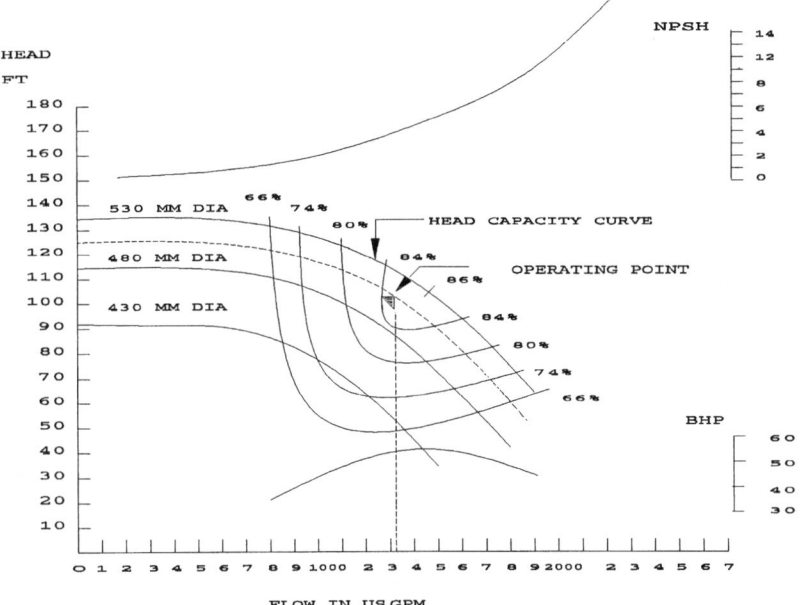

Figure 6.3 Head Capacity Curve With 3% NPSH Curve

48 Practical Introduction to Pumping Technology

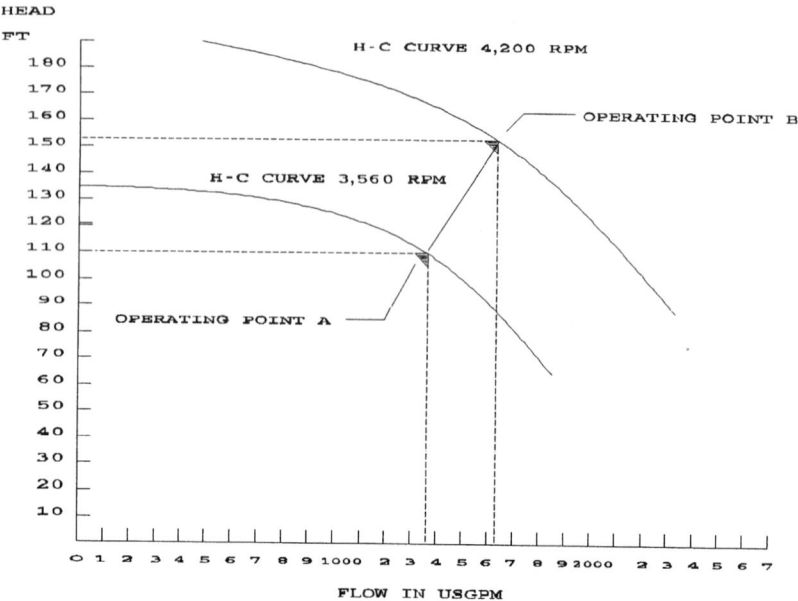

Figure 6.4 H-C Curve With RPM Change

System Curves

When the buyer determines the operating point of a pumping system, he or she plots a system curve (Figure 6.5) overlapping the pump performance curve. The intersection of the system curve and the head capacity curve is the operating point. The system curve shows the actual losses in the discharge piping. The curve does not necessarily start at 0 head and 0 flow. Usually, the end of the discharge line requires a fixed head, for instance the pressure of a pipeline into which the liquid is pumped, or the positive head in a tank. This function appears as a straight line because this head remains the same for all flows, minimum to maximum.

Friction losses in the pipe, losses in fittings and valves, and exit losses cause the losses at the discharge. Often a back-pressure valve, a flow control valve, or a pressure control valve is added downstream of the discharge check and block valves to maintain a set discharge head.

Pumps Operating in Parallel

Because sooner or later any pump will fail, adding a spare pump will prove prudent. You can do this several ways.

The most common system has a full-capacity pump piped in parallel with another full-capacity pump, both of which discharge into the same line (Figure 6.6). Thus, the system has 100 percent spare capacity, also called "one operating, one spare."

Pump Curves   49

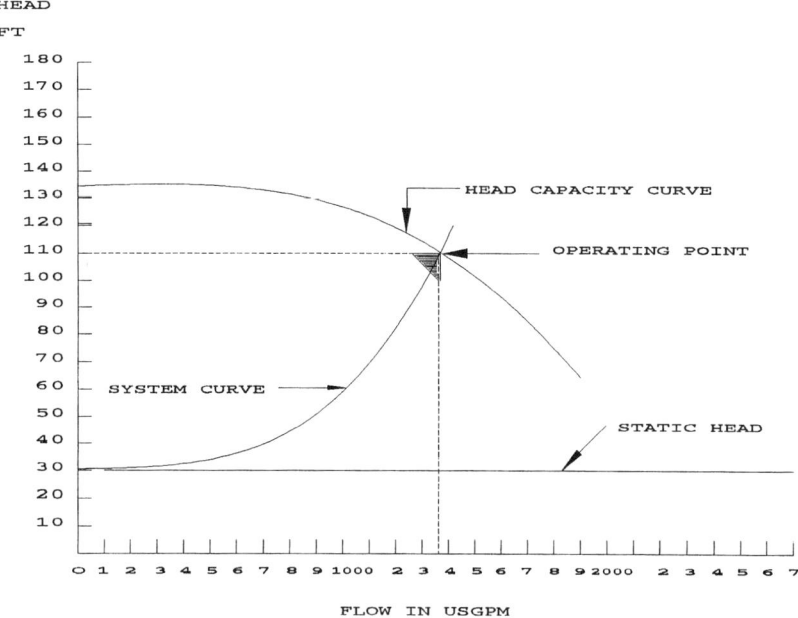

Figure 6.5 Head Capacity and System Curves

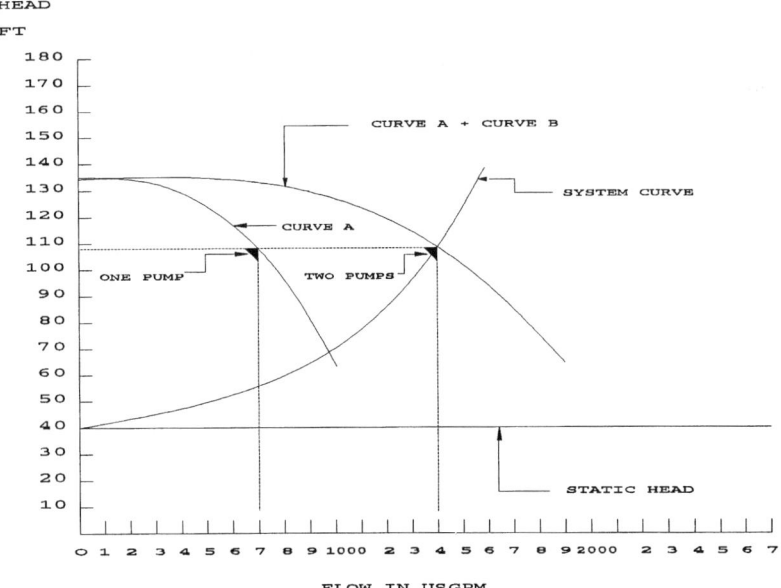

Figure 6.6 H-C Curves of Pumps Operating in Parallel

50 *Practical Introduction to Pumping Technology*

When the user wants more flexibility, he or she may install two pumps in parallel with one spare, also installed in parallel. This is called 50 percent sparing. In this case, all three pumps are identical, with two pumps running to deliver full capacity. If one pump fails, the spare starts either automatically or manually. Figure 6.7 shows a schematic of such an installation. Many other combinations are, of course, also possible.

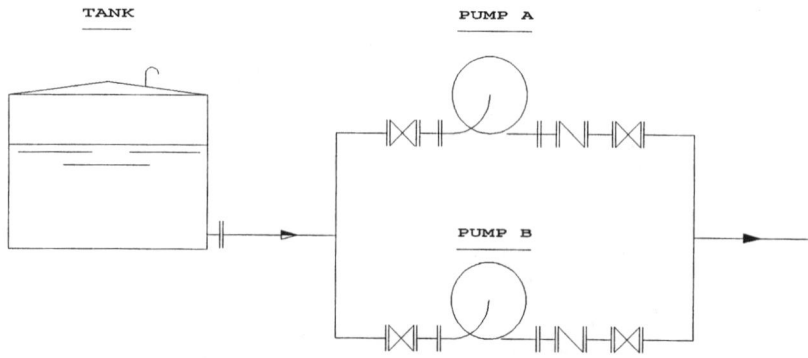

Figure 6.7 Flow Diagram of Two Pumps in Parallel

To calculate the sizes of the pumps needed, the engineer calculates the total flow required by the two pumps. The required flow is 1,400 gpm. The engineer plots the system curve (Figure 6.8). Then he or she selects a pump with the impeller sized so the combined H-C curve of the two pumps will intersect the system curve at a flow rate of 1,400 gpm. At that time, the required head is 108 ft. If at some point Operations wants to run only one pump at the required discharge head, throttling the discharge block valve to bring the operating point back to the desired 108 ft makes this possible.

Often a user buys a pump for a given system, and later the capacity of that pump proves inadequate. When operations requests the capacity be doubled, people commonly make the mistake of purchasing another pump of equal capacity using the same discharge piping configuration. After installing the new pump it becomes apparent—to everybody's chagrin—that the flow has not doubled.

The problem is the system curve has now shifted because of the increased friction losses in the discharge piping system due to higher fluid velocity. Pump A alone operates at 108 ft head at a capacity of 700 gpm. The pump is throttled to achieve this head. When adding another equal pump, Curve A bisects the combined H-C curve at 975 gpm and 128 ft. To achieve twice the original capacity at the same head by adding an additional pump, the discharge configuration needs changing, as represented by System Curve B. The problem may also be solved by using two dissimilar pumps running in parallel (Figure 6.9).

When Pump A runs alone on System Curve A, Q equals 1,200 gpm, and the head is 88 ft. For Pump B, also operating by itself on System Curve A, Q equals 900 gpm against a head of 128 ft. If the two pumps operate in parallel on the same system

Pump Curves 51

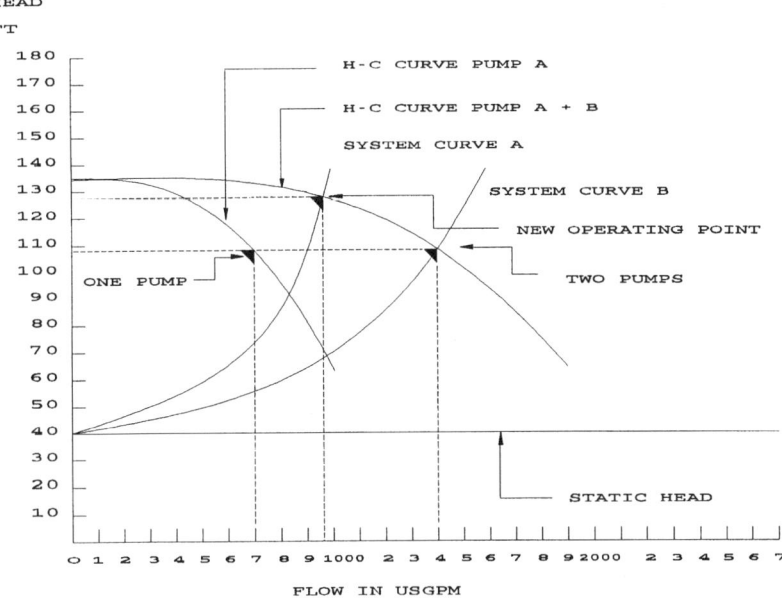

Figure 6.8 System Curves for Pumps Operating in Parallel

curve, Pump A will intersect the system curve before point A. That means Pump A is backed off and Pump B becomes the commanding pump and will deliver the same 900 gpm against a head of 128 ft. To increase the flow, the discharge piping configuration must change. With the new System Curve B, the two pumps running in parallel will deliver 2,200 gpm against a head of 90 ft.

The two pumps can operate correctly only if the H-C curve intersects the system curve on the *AC* portion of the combined H-C curve. The pumps may be throttled, but not further back on their curve than 1,100 gpm, for then Pump A will back off, letting the other pump deliver full capacity. To operate this system, Pump B must start first. After this pump has reached full flow, Pump A may be started.

Pumps Operating in Series

People operate pumps in series to increase the head delivered by a pumping system. The flow diagram (Figure 6.10) shows such an arrangement. Figure 6.11 shows curves of pumps operating in series.

Operating on System Curve A, each pump by itself delivers 800 gpm against a head of 89 ft. The same two pumps in series will deliver 1,250 gpm at a head of 132 ft. The capacity of the two pumps operating in series has increased 56 percent, whereas the head has increased 65 percent.

Operating on System Curve B, each of two pumps delivers 1,300 gpm at ahead of 65 ft. When running the same two pumps in series, the combined capacity equals

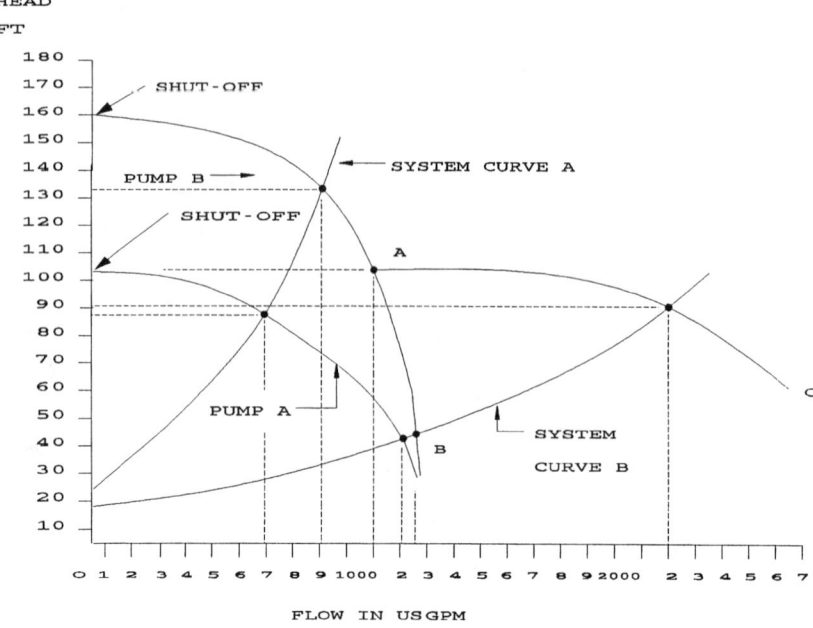

Figure 6.9 Two Dissimilar Pumps Operating in Parallel

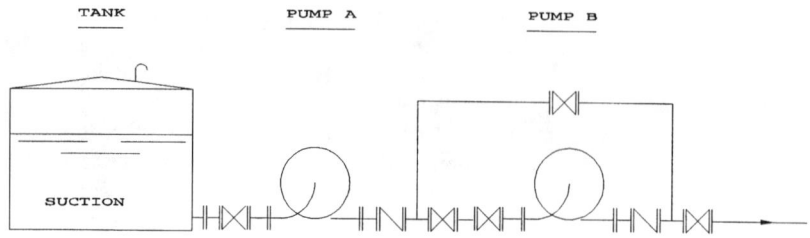

Figure 6.10 Flow Diagram of Two Pumps Operating in Series

1,800 gpm, an increase of 38 percent. The head for the two pumps in series is 92 ft, an increase of 41 ft.

High-pressure, high-volume pumps, such as boiler feed pumps or oil field water injection pumps, require a high NPSH. In general, that NPSHR is not easily available. To improve the conditions, one or several booster pumps in parallel that will deliver the same capacity as the main pump are piped in series with the main pump.

The parameters for the main pump are a capacity of 2,000 gpm against a head of 132 ft, with the NPSHR at 45 ft. The NPSHA available is 9 ft. Therefore, the pump needs an additional 40 ft (36 ft + 10 percent safety or 4 more ft). As seen in Figure

Pump Curves 53

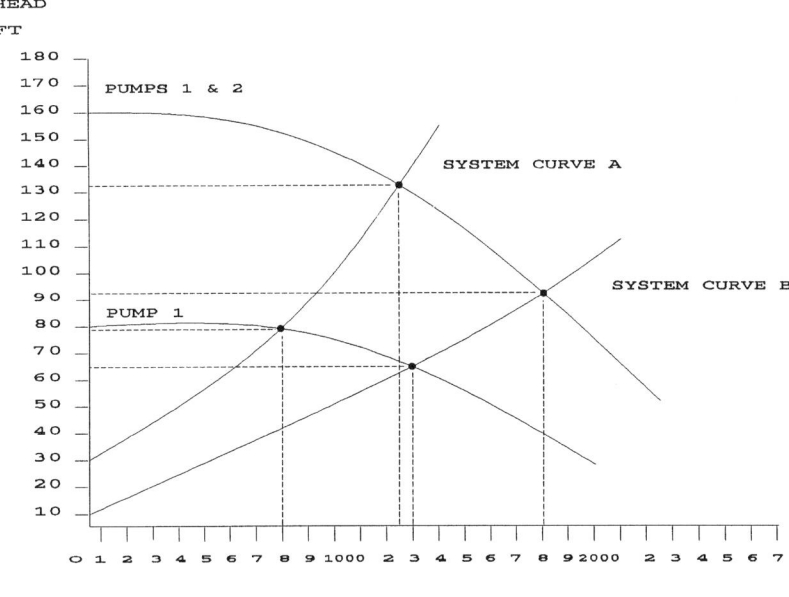

Figure 6.11 Curves of Two Pumps Operating in Series

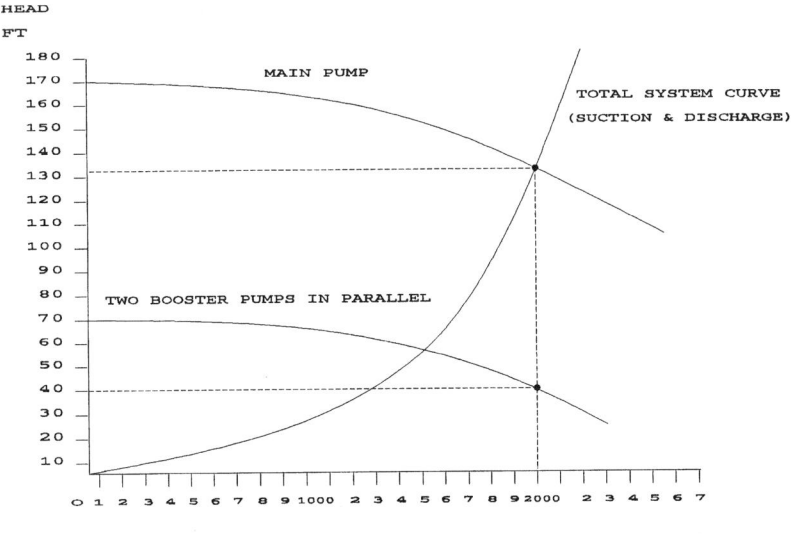

Figure 6.12 System Curves of Main and Booster Pumps

54   Pumping Primer

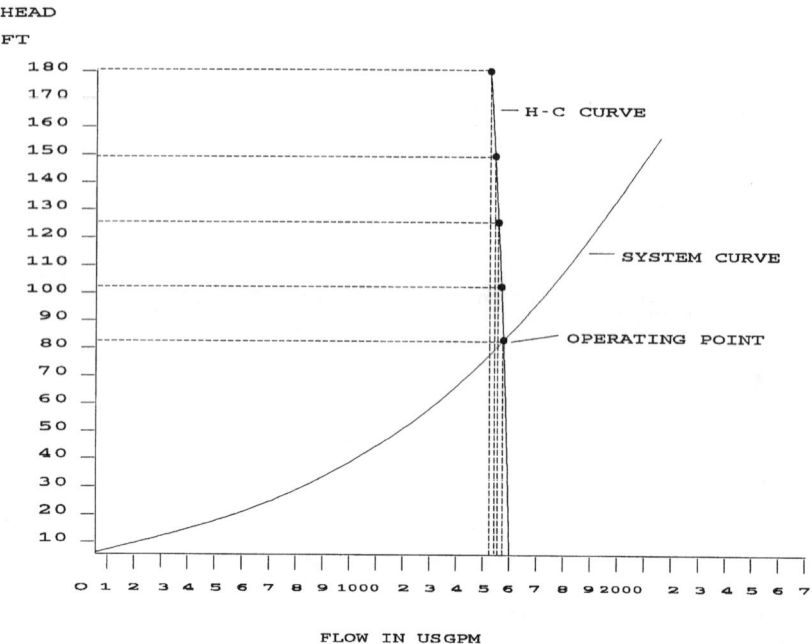

Figure 6.13 H-C Curve for a Positive Displacement Pump

6.12, the two booster pumps are delivering 2,000 gpm into the suction of the main pump, thus boosting the NPSHA of that pump to 49 ft.

Positive Displacement Pump Curves

These pumps have nearly straight, vertical head capacity curves (Figure 6.13). The sizes of the cylinders or other types of cavities as well as the rpm of the crankcase govern the pumps' capacities. Throttling the discharge is not recommended. A little throttling does not reduce the capacity much. Throttling the discharge enough to change the capacity significantly will increase the discharge pressure until the pump pressure may exceed the design pressure.

Chapter 7

Effects of Viscosity on Pump Performance

Imagine viscosity as a liquid's resistance to flow. The viscosity units involve distance and time, and the expression for viscosity is either:

- Dynamic or absolute viscosity
- Kinematic viscosity

Dynamic (Absolute) Viscosity

Referring to the internal resistance of a liquid, dynamic viscosity (μ) is the resistance offered by a fluid or gas to the internal motion of its parts. Poise is the basic unit expressed in mass [g/(sec x cm)]. You can get centipoise, a unit used more commonly with pump applications, by dividing poise by 100 (poise/100).

Kinematic Viscosity

This type of viscosity relates these internal forces to the liquid's specific gravity. The stoke is its unit of measurement, expressed as cm^2/sec. Convert it to the more widely used centistoke by dividing stokes by 100. To convert centipoise to centistoke use:

Centistoke = centipoise × specific gravity

Viscosity Units

Some common units used to measure viscosity include:

- Seconds Saybolt Universal (SSU)
- Seconds Saybolt Furol (SSF)
- Seconds Redwood 1 Standard
- Seconds Redwood 2 Admiralty
- Degrees Engler
- Centipoise
- Centistoke

The first five units listed above have no direct relationship to centistoke or centipoise. Table 7.1 gives some conversion values for the different units.

Table 7.1
Viscosity Conversion Table

Centistokes	SSU	Redwood 1	Engler	SSF
1	31	29	51	
2	33	31	57	
3	36	33	63	
4	39	35	67	
5	42	38	71	
6	45	40	76	
8	52	46	85	
9	55	49	89	
10	59	52	94	
50	231	203	340	26
80	370	324	559	39
100	460	405	677	48
400	1,850	1,620	2,700	188
800	3,700	3,240	5,400	376
1,000	4,600	4,100	6,800	470
3,000	13,900	12,200	20,300	1,400
6,000	27,700	24,300	40,600	2,800
8,000	37,000	32,400	53,800	3,800
10,000	46,200	40,500	67,700	4,700

Because these values may vary with temperature changes, don't consider them completely accurate. One of the most common instruments that measures viscosity is the Saybolt viscosity meter. Other well-known viscosimeters are Redwood and Engler. The Saybolt viscosimeter measures the time a liquid takes to flow through a short, calibrated tube. The flow measure is SSU or SSF. One SSU equals water at 85°F. The following conversion charts are courtesy of the Hydraulic Institute (Figures 7.1 and 7.2).

Industry Preferences

The pump industry prefers to use kinematic or dynamic viscosity at actual pumping temperatures. Common reference temperatures are 100°F and 140°F. When the pumping temperature differs from any of these values, the actual viscosity must be interpolated from product data and/or general viscosity and temperature charts.

Viscosity changes with temperature. As temperature increases, the viscosity of a liquid decreases. A small rise in temperature can lower the viscosity considerably. As the liquid viscosity increases, so do the pump horsepower requirements.

Centrifugal pump manufacturers calculate their performance curves using water as the fluid. When presented with a high viscosity liquid, a manufacturer will superimpose a corrected H-C curve on its regular performance curve. The conversion factor readings on Figures 7.1 and 7.2 are approximate. For exact figures, contact the pump manufacturers.

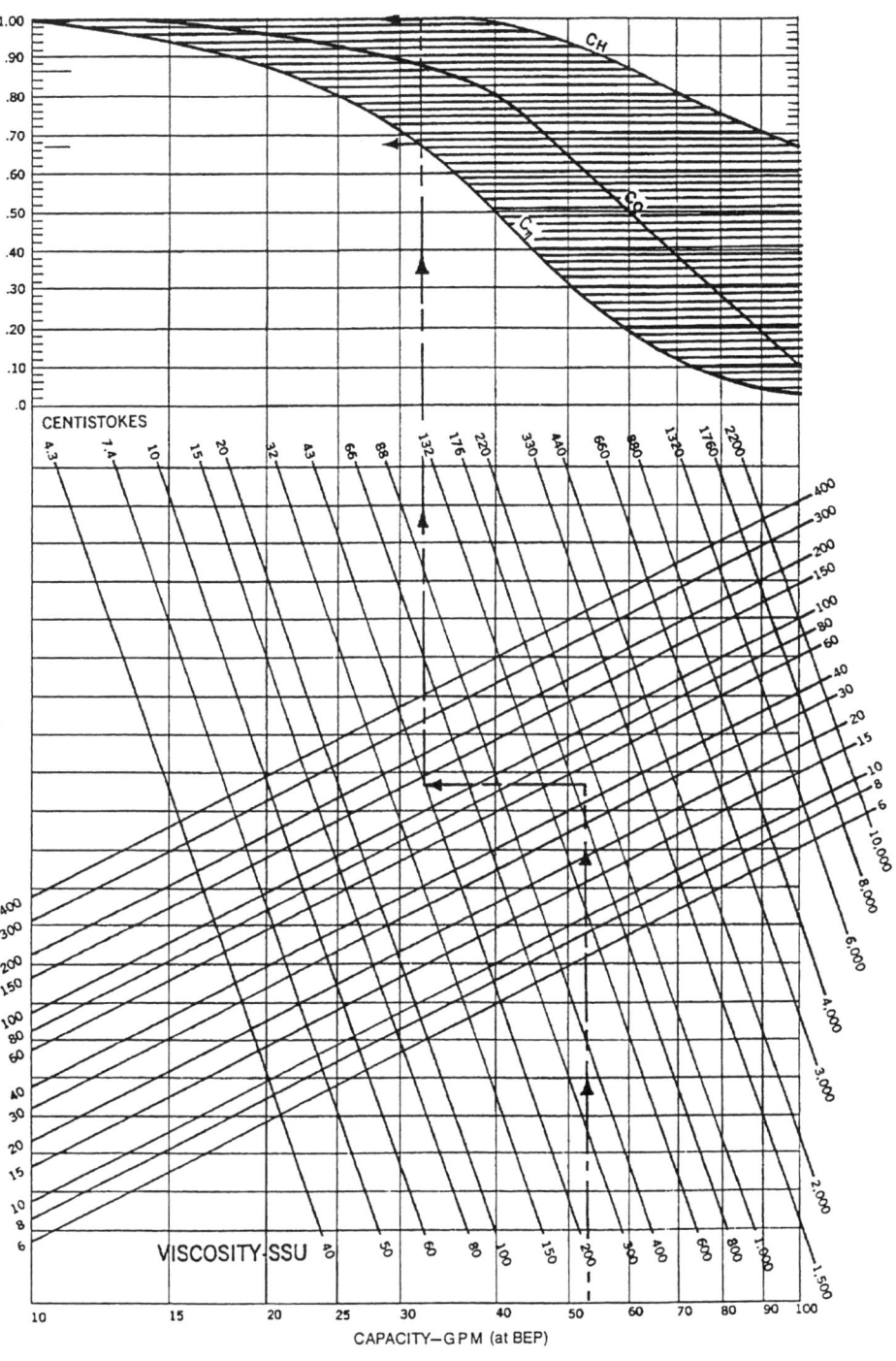

Figure 7.1 Performance Correction Chart for Viscous Liquids

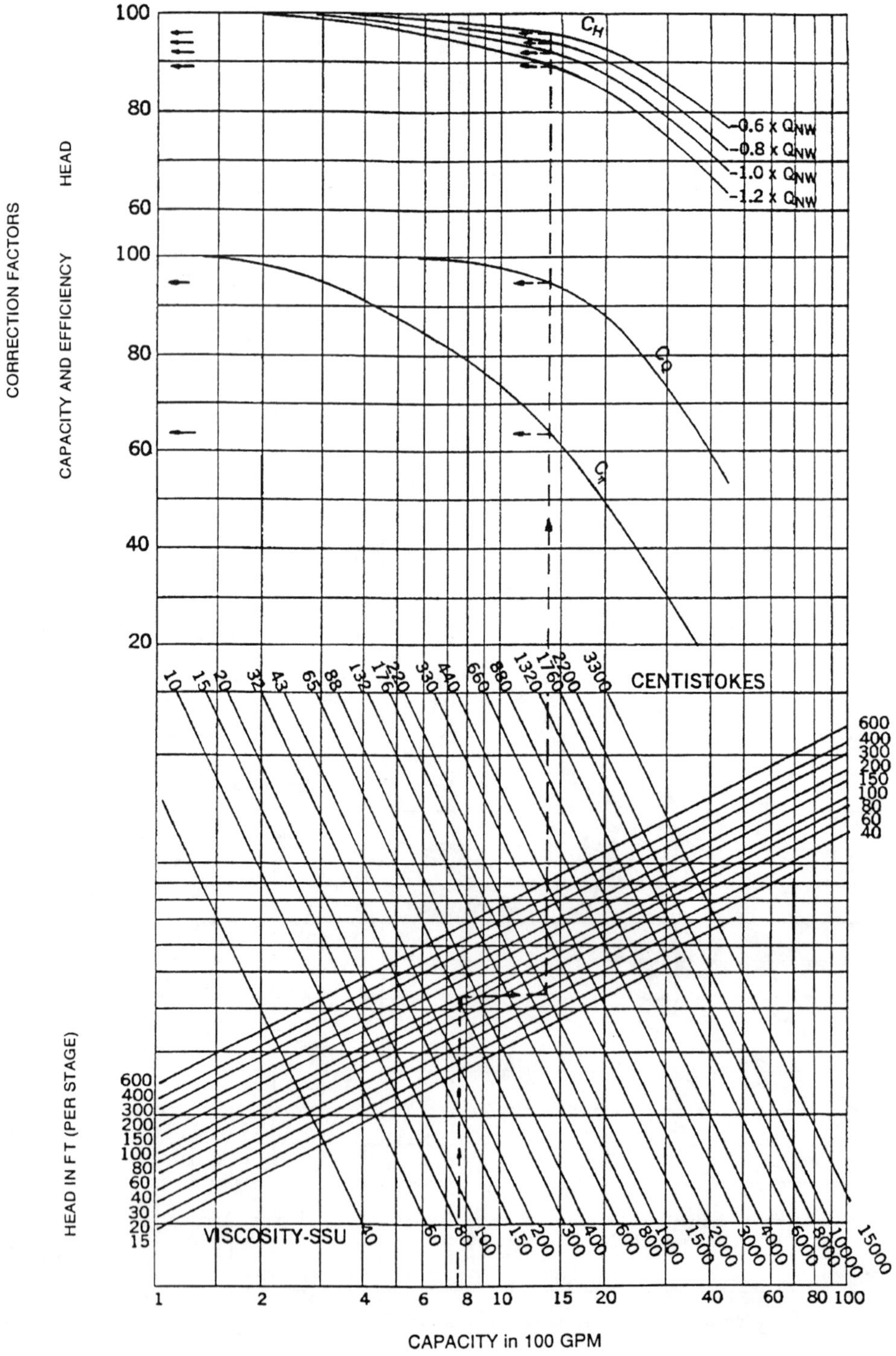

Figure 7.2 Performance Correction Chart for Viscous Liquids

The requirements for a centrifugal pump are as follows:

Fluid = crude oil (California)   (Problem 7.1)
Q = 1,000 gpm
TDH = 150 ft
Viscosity = 660 centistoke
t = 60°F
sp gr = 0.87

On the chart in Figure 7.2, draw a line from 1,000 gpm at the bottom of the chart to the line *H = 150 ft*, then continue horizontally to the right until the line crosses *Viscosity = 660 centistoke*. Continue the line upward and mark the interceptions of Curves C_E, C_Q, and C_H. The correction factors are:

$C_E = 0.48$
$C_H = 0.86$
$C_Q = 0.87$

To find the equivalent water head (H_W) and flow (Q_W), use the following conversions:

H_w = 150/0.86 = 174 ft

Q_w = 1,000/0.87 = 1,149 gpm

The efficiency at BEP is 75 percent. The corrected efficiency is:

E_C (crude) = 0.48 × 75% = 36%

Horsepower requirements for the water are:

(1,000 × 150 × 0.87)/(3,960 × 0.75) = 43.9 BHP

Horsepower requirements for the crude oil are:

(1,149 × 174 × 0.87)/(3,960 × 0.36) = 122 BHP

If the buyer plans to use the original pump for the crude oil, the H-C performance curve is that of Curve C, shown on Figure 7.2. To plot this curve, follow the same procedure as before and multiply the flow, head, and efficiency with the correction factors.

H_c (at rated flow) = 150 × 0.86 = 129 ft

Q_c (at rated flow) = 1,000 × 0.86 = 870 gpm

E_c (at rated flow) = 0.75 × 0.48 = 36%

Calculate some more points, such as minimum flow, 85 percent of rated flow, and 120 percent of the same.

All centrifugal pumps do not perform the same when handling a viscous liquid. Effectiveness varies with each pump's specific speed and physical size. A centrifugal pump with low flow and head requirements can handle only low viscosity fluids. The buyer should verify the pump's capabilities at certain viscosities with the manufacturers. If the efficiency is too low and the power requirement excessive, use a positive displacement pump.

Chapter 8
Vibration

Excessive equipment vibration generally signifies a mechanical malfunction. Therefore, a vibration analysis on both rotating and reciprocating equipment is necessary. As a minimum, these tests shall comply to the following applicable codes and standards:

- *ISO 2372, Mechanical Vibration of Machines*
- *ISO 2373, Mechanical Vibration of Certain Rotating Electrical Machinery*
- *NEMA-MG1-20, Motor and Generator Balance Tolerances*
- *IEC-222, Method of Specifying Auxiliary Equipment for Vibration Measurements*
- *API-610, Centrifugal Pumps*
- *API-611, General-Purpose Steam Turbines*
- *API-613, High-Speed, Special-Purpose Gear Units*
- *API-616, Combustion Gas Turbines*
- *API-670, Noncontacting Vibration and Axial Positioning Monitoring Systems*

If none of these codes apply and the buyer's company does not have vibration testing specifications, the buyer may accept the vendor's standard tests but should add the following paragraph to the pump specification:

> The pump assembly shall run at operating speed until bearing temperatures are stable. A stable condition is when no temperature change greater than 2.5 percent, taken at 5 min intervals, occurs. The pump assembly shall run no less than ½ hr before starting vibration testing. The test shall simulate actual field conditions.

Terms and Definitions

Acceleration: A velocity increase, given as peak G's (gravity values). To convert in./sec^2 to gravity values, multiply by 386. To convert mm/sec^2, multiply by 9,800

Amplitude: The range of the vibration, given as displacement, velocity, or acceleration (see Figure 8.1)

Critical Speed: The speed of a rotating element that falls within the resonance frequency of the element

Displacement: The real motion of a body, given in mils (0.001 in.) or microns (0.001 mm), from peak to peak

Filter In (Filtered): Oscillations a vibration analyzer sorts according to their frequencies

Filter Out (Unfiltered): Unsorted vibrations detected by a vibration analyzer; the largest vibrations a pickup, or sensor, at any position senses

Frequency: Number of vibration cycles expressed in hertz (Hz)

Phase: The location of a vibrating part, with respect to a fixed position. An oscilloscope, an electromagnetic pickup, or a photocell reads the vibrations.

Proximity Probe: An electronic instrument measuring changes in movements

Resonance: A significant vibration amplitude boost that happens when vibration frequencies coincide with the element's normal frequency

RMS Level: A root mean square vibration scale that may identify potentially catastrophic vibrations

Seismic Pickup: A transducer that measures vibration amplitudes

Simple Harmonic Motion: A continuous vibration that has the shape of a sinus curve

Velocity: The maximum speed of a point on a vibrating body, given in in./sec or mm/sec

Vibration: An oscillating motion produced by a force and estimated by an amplitude, a frequency, and a phase angle

Testing Procedures

The pump buyer doesn't usually witness vibration tests on smaller pumps. Larger pumps (20 hp and larger) warrant witnessed tests by the buyer or his or her represen-

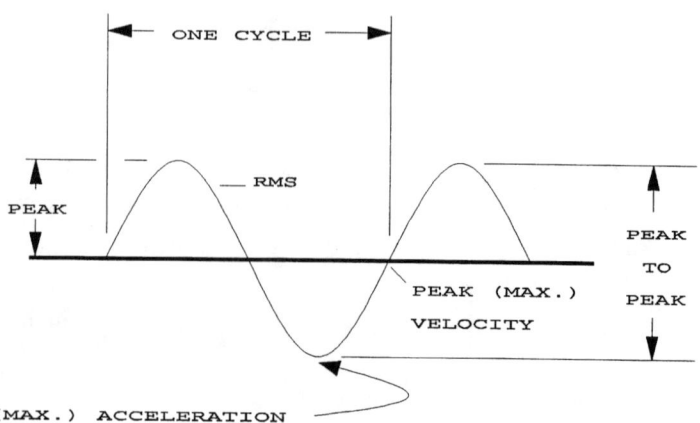

Figure 8.1 Amplitude Measurements

tative. The vibration analysis may be made part of the mechanical acceptance test, if both buyer and vendor agree to do so.

The vendor shall show both first and second critical speeds on test records, including the shaft speeds on any gears included in the pump package. The inspector shall log the vibration levels from the horizontal, axial, and vertical positions on all reachable bearings.

The vibration analysis shall highlight the vibration levels and frequencies at the following speeds:

- Design speed
- Design speed × 2
- 50 percent of design speed
- 10 percent above design speed

Vibration Limits

The following tables show some suggested acceptable vibration amplitudes for pumps and drivers:

Table 8.1
Electric Motors

NEMA Approximate Frame Size	Horsepower Range	Speed (rpm)	Displacement (mils, peak–peak)	Velocity (in./sec)
143T– 184T	0.5–2.0	3,600 & less	0.75	0.15
213R–286T	2.0–30.0	3,600 & less	1.00	0.20
324T–445T	40.0–350.0	3,600 & less	1.00	0.20
Large motors	350.0 & above	3,600 & less	1.00	0.20

On small motors, make readings on bearing housing.
Readings on large motors with sleeve bearings shall be made on the motor shaft.

Table 8.2
Centrifugal Pumps

| Speed (rpm) | Displacement (mils, peak–peak) Bearing || Velocity (in./sec) Bearing ||
	Antifriction	Sleeve	Antifriction	Sleeve
1,800 & less	3.0		0.27	
1,801–4,500	2.0	2.0	0.30	0.30
4,501–6,000		1.5		0.33
6,001 & higher		1.0		0.35

On pumps with antifriction bearings, measure the vibrations on the bearing housing close to the centerline.
Measure vibrations for sleeve bearings on the shaft.

Table 8.3
Steam Turbines, High-Speed Parallel Gears, Epicyclic Gears, and Rotary Pumps

Speed (rpm)	Displacement (mils, peak–peak)	Velocity (in./sec, peak)
5,000 & less	7.0	0.25
501–1,000	5.0	0.27
1,001–3,000	3.0	0.30
3,001–4,500	2.0	0.31
4,501–6,000	1.5	0.32
6,001–8,000	1.0	0.33

For acceptance test, use peak velocity readings.

Table 8.4
Low-Speed and Bevel Gears

Speed (rpm)	Displacement (mils, peak–peak)	Velocity (in./sec, peak)
100 & less	14	0.25
101–200	12	0.26
201–300	10	0.27
301–400	8	0.28
401–500	7	0.29
501–600	6	0.30
601–700	5	0.30

For acceptance test, use peak velocity readings.

Table 8.5
Aircraft Derivative Gas Turbines

Speed (rpm)	Displacement (mils, peak–peak)	Velocity (in./sec, peak)
1,000 & less	8.00	0.45
1,001–4,000	4.00	0.50
4,001–6,000	2.00	0.52
6,001–8,000	1.50	0.52
8,001–12,000	1.00	0.50
12,001–18,000	0.75	0.45
18,001–25,000	0.50	0.40
25,001 & higher	0.30	0.30

For acceptance test, use peak velocity readings.

Table 8.6
Industrial Gas Turbines

Speed (rpm)	Displacement (mils, peak–peak)	Velocity (in./sec, peak)
1,000 & less	5.00	0.27
1,001–4,000	2.50	0.32
4,001–6,000	1.50	0.33
6,001–8,000	1.00	0.34
8,001–12,000	0.70	0.33
12,001–18,000	0.60	0.32
18,001–25,000	0.35	0.30
25,001 & higher	0.25	0.27

For acceptance test, use peak velocity readings.

Table 8.7
Reciprocating Engines

Speed (rpm)	Displacement (mils, peak–peak)	Velocity (in./sec, peak)
50–500	10.0	0.27
501–1,000	7.0	0.32
1,001–1,500	5.0	0.32
1,501–2,000	4.0	0.33
2,000 & higher	3.0	0.33

For acceptance test, use peak velocity readings.

Induced Piping Vibrations

Restrict piping vibrations caused by a pump package as follows:

- Vibrations on all piping that is attached to the pump package and/or is internal shall have the same restrictions as the package itself.
- Vibration limits on piping outside the pump package no more than 10 ft away from discharge and suction flanges shall be the same as the limits on the pump package.
- Piping upstream or downstream of the first pipe support shall be less than twice the acceptable amplitude limits of the pump package. If vibrations exceed this, you'll need additional pipe supports to lessen the vibrations.

Chapter 9
Net Positive Suction Head (NPSH)

Definition

The term *net positive suction head* confuses many people. It differs from both suction head and suction pressure. For instance, when an impeller in a centrifugal pump spins, the motion creates a partial vacuum in the impeller eye. The NPSH is the height of a column of liquid that will fill this partial vacuum without allowing the liquid's vapor pressure to drop below its flash point. In other words, this is the NPSH required (NPSHR) for the pump to function properly.

The Hydraulic Institute defines NPSH as "the total suction head in feet of liquid absolute determined at the suction nozzle and referred to datum less the vapor pressure of the liquid in feet absolute." This defines the NPSH available (NPSHA) for the pump. Considering only these two parameters, a pump will run satisfactorily if the NPSHA equals or exceeds the NPSHR. Because most pumps are not high-precision machinery, most authorities recommend the NPSHA be at least 2 ft absolute or 10 percent larger than the NPSHR, whichever number is larger.

Calculations can determine the NPSHR of a particular pump design. Usually, though, pump manufacturers test the pumps to find the NPSHR. The test basically consists of finding where the liquid's vapor pressure exceeds the NPSHA when the liquid enters the eye of the impeller, causing the pump to cavitate.

The NPSHR test consists of running a pump at constant speed while throttling the pump suction line until the discharge pressure drops. At that point, the pump starts to cavitate. Incipient cavitation happens as soon as the discharge pressure begins to drop, called 0 percent head drop. Because this point is difficult to mark, the NPSHR is recorded at a 3 percent head drop. Points are plotted from 0 capacity to end-of-the-curve capacity. The Hydraulic Institute recommends this method.

NPSH Calculations

The following examples show how to determine the NPSHA for different pumping systems.

Net Positive Suction Head (NPSH) **67**

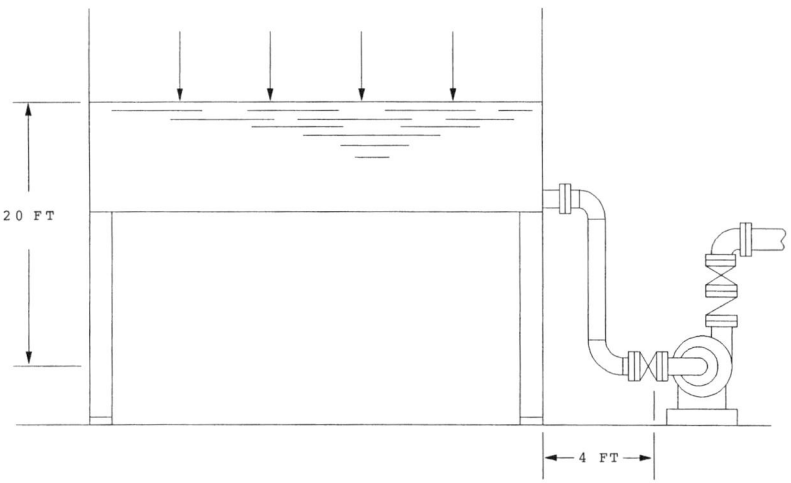

Figure 9.1 Open Atmospheric Tank

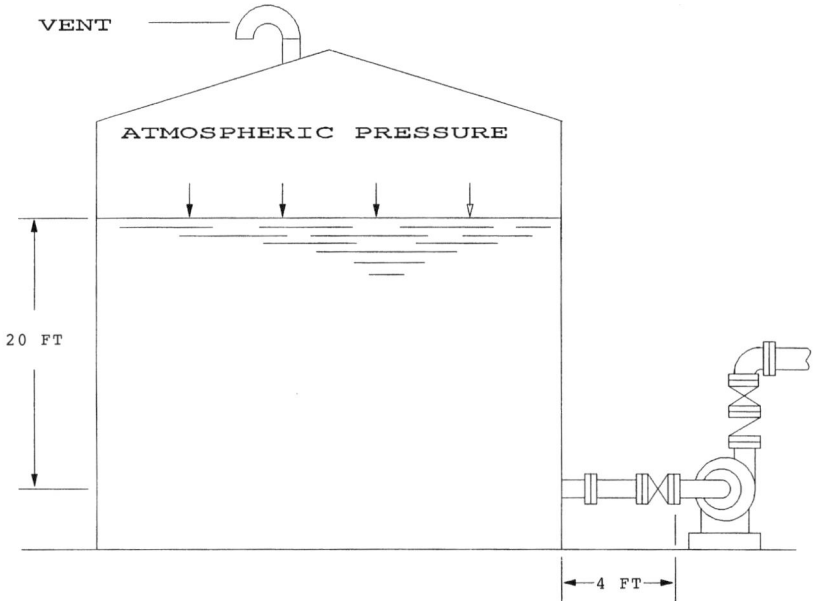

Figure 9.2 Roofed Water Storage Tank

Atmospheric Tank

This may be an open-top tank or a municipal water storage tank with a roof and correctly sized vent (Figures 9.1 and 9.2).

The formula for calculating NPSHA is:

$$\text{NPSHA} = P_a + h - P_v - h_e - h_f$$

where:

P_v = vapor pressure in absolute of liquid at given temperature
P_a = atmospheric pressure in absolute or pressure of gases against surface of liquid
h = weight of liquid column from surface of liquid to center of pump suction nozzle in feet absolute
h_e = entrance losses in feet absolute
h_f = friction losses in suction line in feet absolute

Given the following, find the NPSHA.

    Liquid = water   (Problem 9.1)
      t = 60°F
    sp gr = 1.0
      P_v = 0.256 psia (0.6 ft)
      h_e = 0.4 ft
      h = 20 ft.
      h_f = 2 ft
      P_a = 14.7 psia (34 ft)

In this case:

NPSHA = 34 ft + 20 ft − 0.6 ft − 0.4 ft − 2 ft

NPSHA = 51 ft

Suction Lift from Open Reservoir (Figure 9.3)

Find the NPSHA, where:

    Liquid = water   (Problem 9.2)
      t = 60°F
    sp gr = 1.0
      P_v = 0.256 psia (0.6 ft)
      Q = 100 gpm
      h_e = 0.4 ft
      h_f = 2 ft
      h = −15 ft
      P_a = 14.7 psia (34 ft)

Net Positive Suction Head (NPSH)

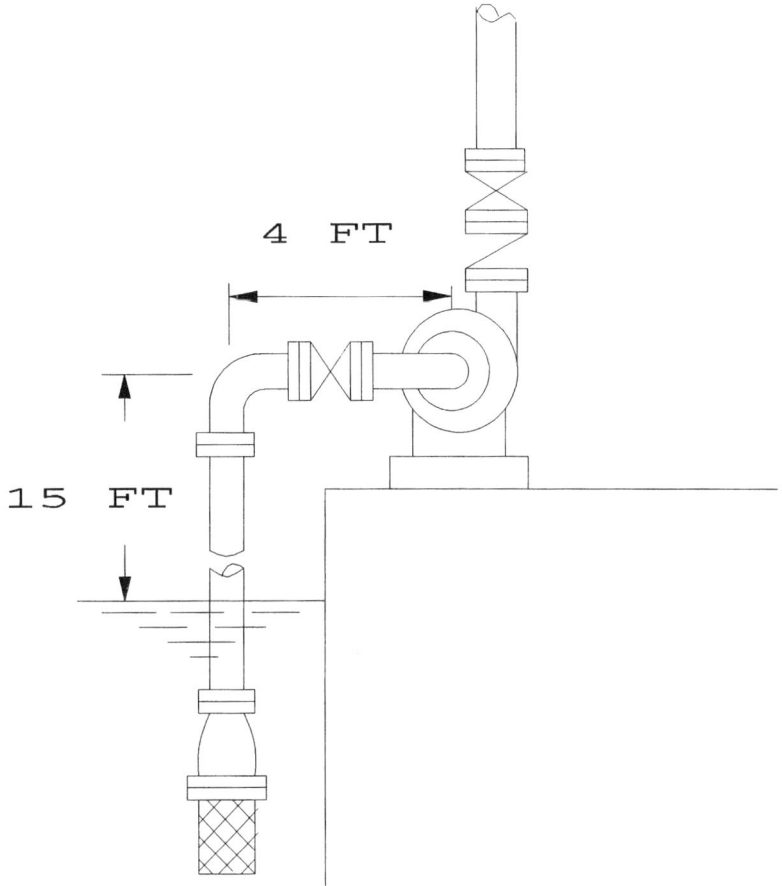

Figure 9.3 Suction Lift From Open Reservoir

NPSHA = 34 ft − 15 ft − 0.6 ft − 0.4 ft − 2 ft

NPSHA = 16 ft.

Pressure Vessel

In a pressure vessel, the liquid is in equilibrium (Figure 9.4). This means the liquid's vapor pressure equals the pressure of the gases upon the liquid. In this case, the NPSHA equals the height of liquid in the column, from the surface of the liquid in the vessel to the center of the suction nozzle of the pump, minus friction losses.

Find the NPSHA, where:

70   *Practical Introduction to Pumping Technology*

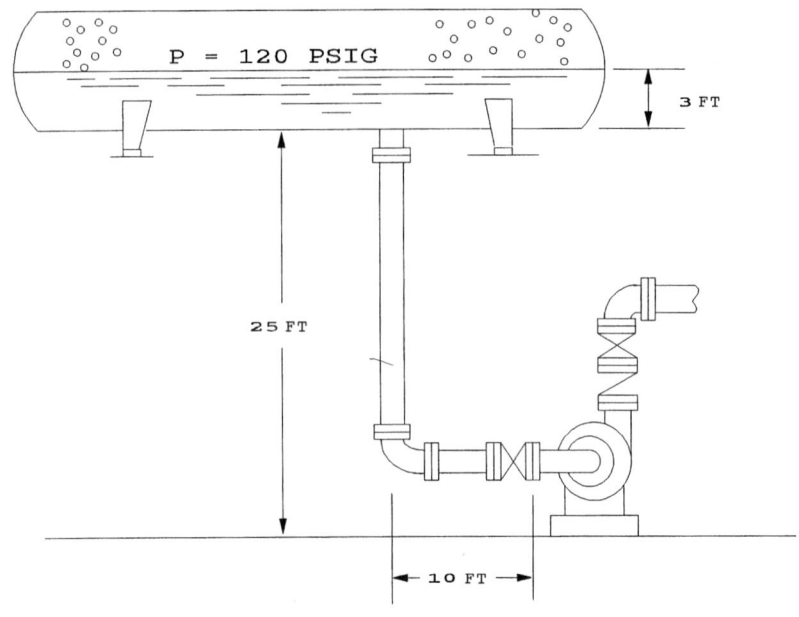

Figure 9.4 Pressure Vessel

Liquid = wet crude   (Problem 9.3)
t = 75°F
sp gr = 0.85
Operating pressure P_a = 52.2 psia (141.9 ft)
P_v = 52.2 psia (141.9 ft)
h = 5.6 ft
h_e = 0.4 ft
h_f = 2 ft

NPSHA = 141.9 ft + 5.6 ft − 141.9 ft − 0.4 ft − 2 ft

NPSHA = 3.2 ft

Crude Oil Tank With Nitrogen Blanket

This tank, although similar to the one shown in Figure 9.2, also has a vacuum breaker and a safety relief valve (PSV). Pressure of the nitrogen blanket is 0.5 psia (1.4 ft). Find the NPSHA, where:

Liquid = lake water   (Problem 9.4)
t = 68°F
sp gr = 1.0
P_v = 0.783 psia

h = 15 ft
h_e = 1.6 ft
h_f = 3.2 ft
P_a = 14.7 psia (34 ft)

NPSHA = 34 ft − 0.783 − 15 ft − 3.2 ft − 1.6 ft

NPSHA = 13.4 ft

Additional Requirements

For pumps with low flow and moderate heads, the NPSH values supplied by pump manufacturers on their H-C curves will suffice (Figure 9.5).

Sometimes a buyer will request a 0 percent head drop NPSH curve from a pump manufacturer (Figure 9.6). The Hydraulic Institute publishes a method to calculate this head drop.

NPSH requirements grow as the capacity and discharge pressures increase. Large, high-discharge-pressure pumps, such as boiler feed, shipping, and water injection pumps, that require high NPSH values warrant a closer look. Buyers often require high impeller life from such pumps. One of the largest oil producing companies in the Middle East requires a 40,000-hr impeller life for its larger pumps.

As the capacity of pump models increases, the shape of the NPSH curve changes. Instead of exponentially rising from shutoff flow to end-of-capacity flow, the curve takes on the shape of an inverted parabola.

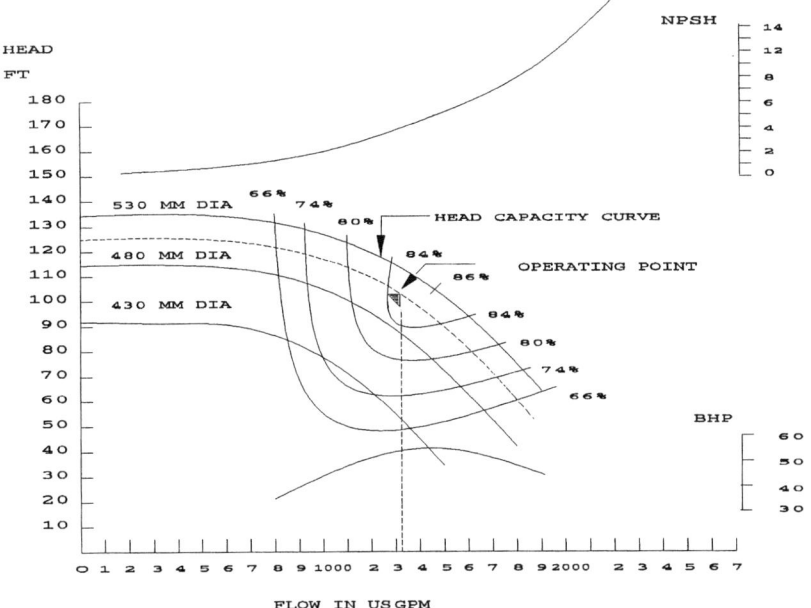

Figure 9.5 Head Capacity Curve With 3% NPSH Curve

72 *Practical Introduction to Pumping Technology*

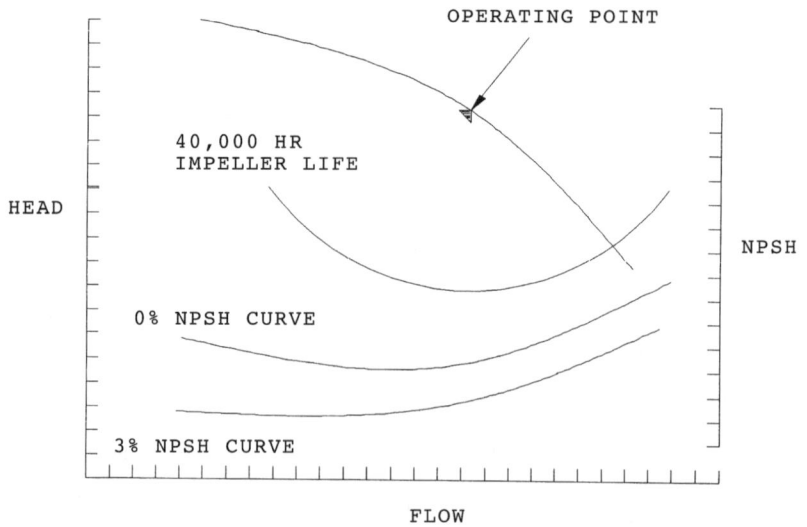

Figure 9.6 0% and 3% Head Drop NPSH Curves

Pump manufacturers use flow visualization tests, both actual tests and computer modeling, to design the geometry of the pumps and impellers. While throttling the pump's suction, the technicians observe the formation of bubbles around the impeller eye. They change the vane angle, vane profile, and size of the impeller eye until they achieve an optimum configuration.

Both the pump material and the liquid influence the NPSH considerably. A type 316 stainless steel impeller will last longer than a brass one. Cold water has a higher gas-to-liquid ratio. At a pressure drop, more gas is released, and the pump cavitates at lower NPSH values. In general, smooth casing and impeller surfaces create less turbulence and thus improve NPSH. As a rule, discard pumps where the ratio NPSHA/NPSHR drops below 1.2.

The lowest NPSHR occurs at the shockless entrance (Q_{sw}) of the liquid. This means the liquid enters the impeller eye without prerotation. Usually pump manufacturers design their impellers so the shockless entrance will be at approximately 105 percent of the BEP to compensate for ring wear, which in time will bring the Q_{sw} back to the pump's BEP.

When a pump operates at its BEP, the liquid's angle of attack is tangential to the impeller vanes. As flow increases, this incidence value becomes increasingly oblique, until at a particular value the flow separates from the impeller, and the impeller stalls. As a result, flow recirculates and performance drops. Local velocities also increase because of stalled, dead water, which may cause the pump to cavitate. To achieve low NPSHR at BEP and higher flow, some pump manufacturers

Net Positive Suction Head (NPSH)

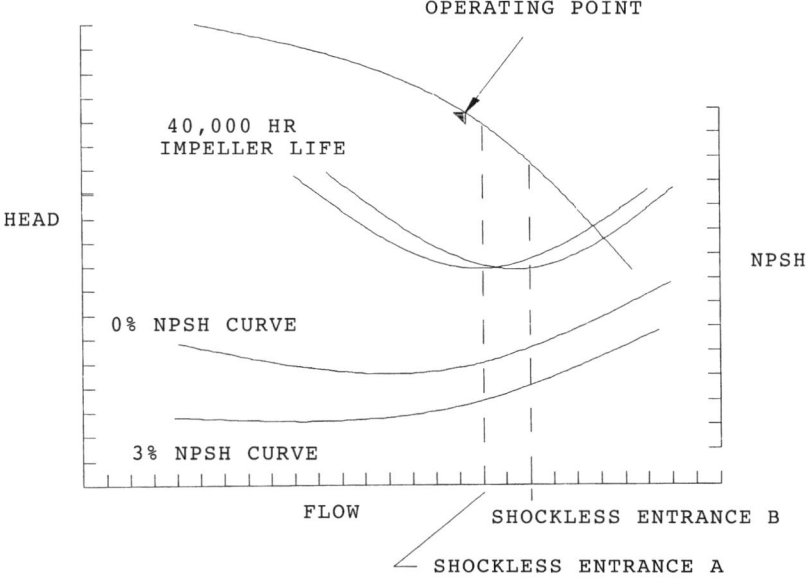

Figure 9.7 Improved NPSH Curves

design their impellers so Q_{sw} occurs at higher flows than the usual 105 percent (see Figure 9.7).

Manufacturers achieve a higher Q_{sw} by increasing vane angle at the inlet side. However, high NPSH requirements at low flow offset this improvement of NPSHR at BEP and higher flows. Although not evident on a 3 percent head drop curve, it shows up on a NPSHR curve calculated for high impeller life.

NPSHR, a function of pump design, relates to the suction capabilities of the first-stage impeller. An impeller with a large eye will require less NPSH but will sacrifice efficiency and exhibit unstable operation at low and/or high flow. The suction specific speed (S) defines the impeller inlet geometry.

A small impeller eye area makes for a low S impeller. High S pumps require high flows to prevent cavitation in the impeller eye. Another problem with a high S impeller is fluid separation and recirculation may occur that will cause cavitation around the impeller eye. High S pumps (11,000 and higher) often demonstrate instability at other than BEP flows. High S pumps' minimum flow requirements exceed that of low S pumps.

Chapter 10
Pump Shaft Sealing

Centrifugal and rotary positive displacement pumps have some sealing arrangements to keep pump fluid from leaking where the pump shaft penetrates the casing. These sealing arrangements may be packed glands or mechanical face seals. Seals also keep foreign material from entering the pump or lubricant from leaking out of bearings and transmissions. Reciprocating pumps have soft packings, molded elastomer rings, piston rings, and metal bushings. Most centrifugal and rotary pumps still have packings, but with mechanical seals becoming more reliable and economical, companies use mechanical seals even in water services.

Packed Glands

Originally, a piece of cotton or hemp rope flushed by the pump fluid served as the pump packing. People in many parts of the world still use asbestos saturated with binders and lubricants, but the potential danger inherent in the material limits its use in the industrialized nations. Modern packings are:

- Braided graphite
- Expanded graphite in preshaped rings or in rolls
- Braided carbon fiber
- Polytetrafluoroethylene (PTFE) fiber
- Braided or PTFE impregnated with graphite
- Extruded PTFE mixed with graphite
- Braided aramide fiber coated with PTFE

A packing gland (Figure 10.1) works through compression. The rings are compressed with enough force to keep the pump fluid from leaking through the shaft. Even though the packing rings are either self-lubricated or grease lubricated, the pumped fluid adds to the lubrication and also cools the packing. Heat breaks down packings. To avoid heat buildup, the packing gland should allow some fluid to leak through. Too tight a gland will destroy the packing by heat deterioration. A lantern ring placed in the middle of the packing will add cooling fluid. Addition of a clean flushing fluid through the lantern ring is desirable when the pumped fluid is abrasive.

Abrasive fluid or too tight a packing gland will eventually scour the part of the shaft beneath the packing. To prolong the shaft life, manufacturers install replaceable shaft sleeves where the shaft comes into contact with the packing.

The advantages of packings over mechanical seals are that the packings save money initially and are easier to install and replace.

Pump Shaft Sealing

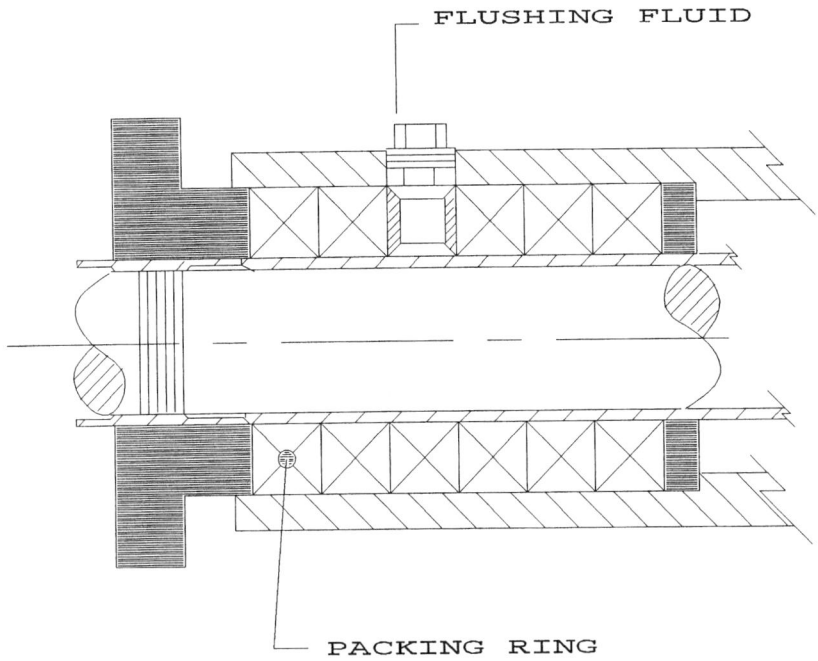

Figure 10.1 Packed Gland Configuration

Mechanical Face Seals

These have been around since the beginning of the twentieth century. The first mechanical seals (Figure 10.2) simply consisted of a shaft collar turning directly against a machined part of the pump casing.

Lack of suitable material and production techniques prevented the seals from being accepted by industry. Later the automobile industry began using a rubber V-ring based on the same principle as the original mechanical seal. With the development of synthetic elastomers and high-quality alloy steel, the chemical industry soon followed suit. Now mechanical face seals (Figure 10.3) are common in all industries. Plants turn them out by the millions, in small batches, or individually, to a buyer's specification.

Processes handling liquids as diverse as sludge, acids, and hydrocarbons—with or without hydrogen sulfite, beer, plastic, and poisonous gas and liquids—use mechanical seals. In the 1940s, maximum seal pressure limits reached about 300 psig with shaft speeds not exceeding 35 ft/sec, but modern seals can tolerate more than 3,000 psig and speeds over 40,000 rpm.

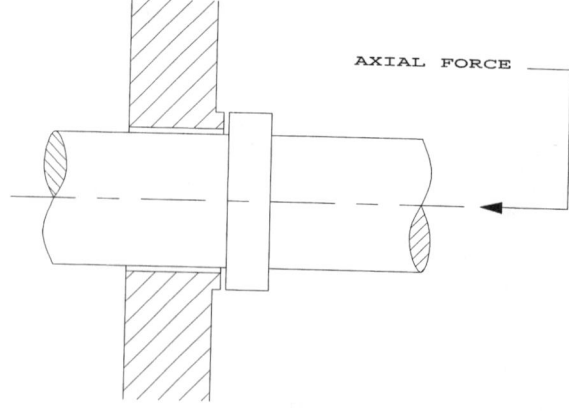

Figure 10.2 Early Face Seal

General Design

Mechanical face seals may have either an axial force pressing a precision-lapped floating ring face against a fixed counterpart, or a fixed ring pressing against a floating ring. Springs or bellows press the faces together. O-rings, V-rings, U-cups, or other types keep the fluids from leaking along the surface of the shaft.

Today almost all mechanical face seals are cartridge types, which means the seals arrive at the customers already assembled. Exceptions include overhung and very large pumps, whose weight may limit their use.

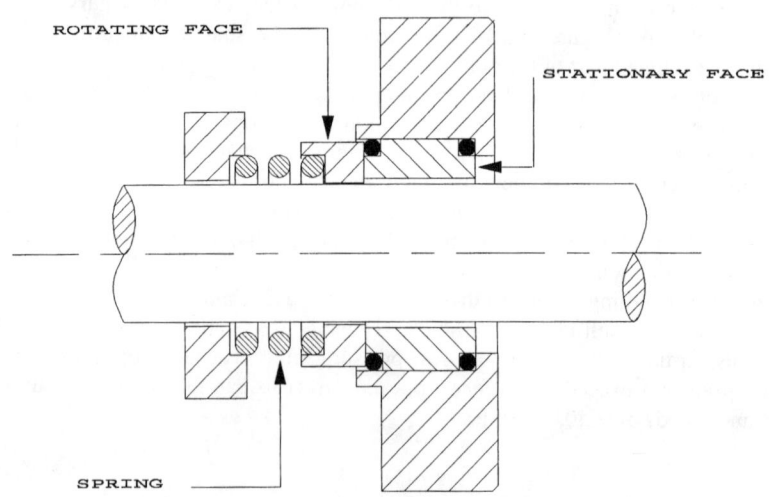

Figure 10.3 Mechanical Face Seal

Mechanical seals for centrifugal and rotary positive displacement pumps break down into various categories.

Internal Seal Arrangements

These arrangements (Figure 10.4) conduct the leakage radially inward. Both internal and external seals may have static or rotating floating rings. The great majority of centrifugal pumps using mechanical seals has an internal seal arrangement.

External Seal Arrangements

These seals (Figure 10.5) conduct the seal leakage outward. The arrangement favors highly corrosive liquids and seals rotating at more than 4,000 rpm.

Balanced and Unbalanced Seals

Balanced and unbalanced seals (Figures 10.6 and 10.7, respectively) differ according to the hydraulic loading of the seal faces. The ratio (A) of the hydraulic recess area (A_h) divided by the interface area (A_i), or $A = A_h/A_i$, determines whether the seal is balanced or unbalanced. When A is smaller than 1, the seal is balanced, and when it exceeds 1, unbalanced. Mostly, the area ratio for unbalanced seals is 1.1 and 1.2, whereas the ratio for balanced seals may vary from 0.6 to 0.9. The risk of face seizing diminishes with lower area ratios, but the possibility of face separation increases.

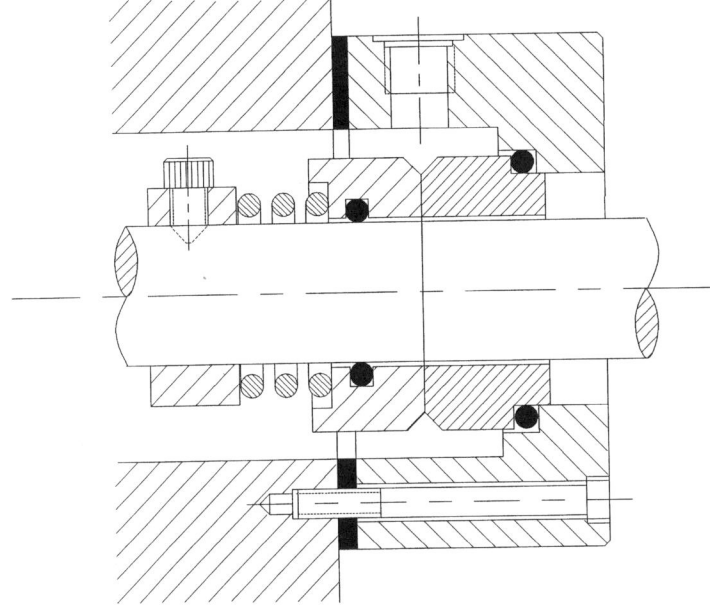

Figure 10.4 Internal Seal Arrangement

78 *Practical Introduction to Pumping Technology*

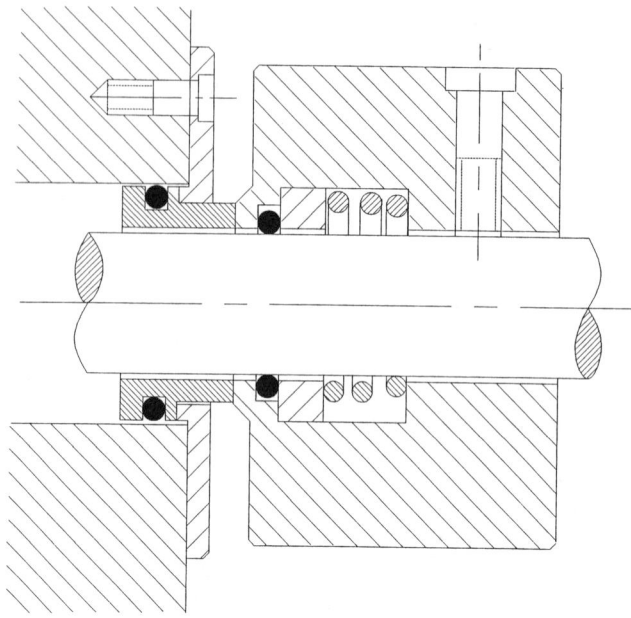

Figure 10.5 External Seal Arrangement

To determine when to use a balanced seal, apply the PV formula, which is the product of the sealing pressure in the stuffing box (P) and the peripheral velocity of the mechanical seal (V), measured at the mean diameter of the seal faces. The following formulas will determine P and V:

P = suction pressure + 0.25 × (discharge pressure − suction pressure)

For vertical turbine pumps, P equals the discharge pressure.

Medium diameter (D_m) = outside diameter (OD) + inside diameter (ID)/2

(*diameters in inches*)

V = D_m × rpm/(3.82 ft/min)

Use 1.5 times the shaft diameter for D_m when you don't know this figure. As a rule of thumb, avoid using unbalanced seals for PVs less than 125,000 psig ft/min and for liquids with a specific gravity less than 0.7.

Basic Seal Types

Single Seals

Single unbalanced or balanced seals are adequate for most services. The pumped liquid may wash the seal faces, or the flush fluid may provide an outside source of

Pump Shaft Sealing 79

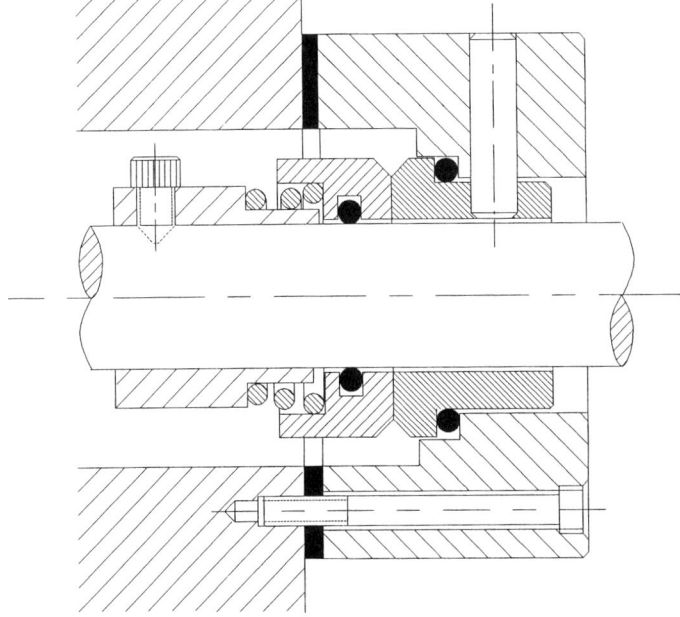

Figure 10.6 Balanced Mechanical Seal

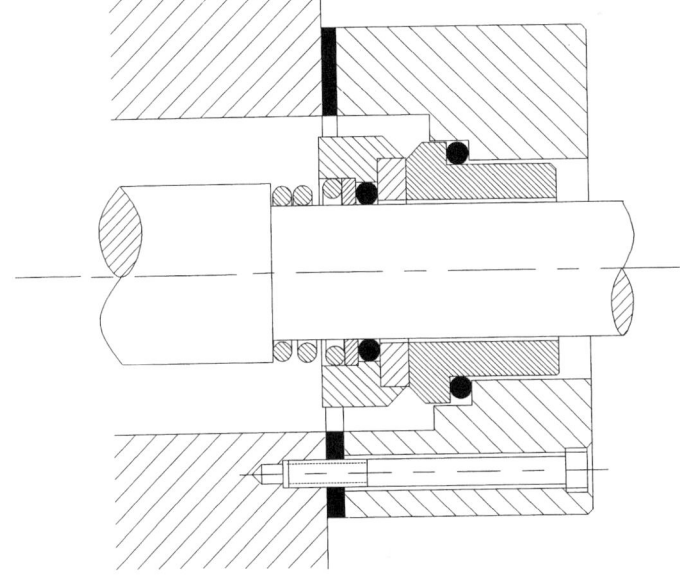

Figure 10.7 Unbalanced Mechanical Seal

clean liquid. Placing a bronze, close-clearance carbon or Teflon™ auxiliary bushing at the seal's atmospheric side gives added protection and provides a place for leak detection equipment. Except for water service, the radial clearances at the atmospheric side of the seal preferably should not measure less than ⅛ in. Some companies recommend an auxiliary gas seal for certain applications.

Tandem Seals

In this type (Figure 10.8), two single mechanical seals occur in series. The pumped liquid flushes the faces of the inner seal, whereas a clean outside buffer fluid flushes the secondary seal faces. The following conditions call for tandem seals:

- Lack of adequate quench fluid, with carbon forming at the seal's atmospheric side or with the pumped liquid crystallizing at the same spot
- Presence of toxic liquid that will vaporize at atmospheric conditions, causing fire hazards
- A need to contain leakage from the primary seal and to delay leakage in case of a catastrophic failure of the primary seal
- Higher stuffing-box pressures than a single mechanical seal can tolerate

Double Mechanical Seals

These seals (Figure 10.9) are placed face to face, with both faces flushed by the same buffer fluid from an outside source. It is desirable to design the seal at the product side so the product fluid cannot enter the seal if the buffer fluid is shutoff. The following conditions call for double mechanical seals:

- The difference between the stuffing-box pressure and the pumped liquid's vapor pressure is less than 25 psi.
- The pumped fluid is abrasive or highly corrosive, and there is no adequate flush liquid.

Face Material

Metal seal faces are not practical because of the high friction coefficient, which creates the risk of seizure and overheating during emergency conditions. In hydrocarbon service, steel with a high graphite content runs against cast or sintered metals. However, carbon, plastic, and ceramic generally are run against metals, metal oxides, and carbides. If the material for the mating faces matches, one of the faces must be of a different grade of material.

Mechanical seals are code classified, usually according to *API Standard 610*, though many manufacturers use their own codes. Appendix 8 shows suggested materials for various liquids.

For instance, an oil company may have the following specification for wet crude, pumped in several different conditions:

- Use code BSAFJ. Flush primary seal with dry crude from an outside source. The auxiliary gas seal shall be BSPFX, where X shall be carbon against sapphire.
- Use code BSTFM if wet crude flush is not available. Flush shall be clean water.

Pump Shaft Sealing 81

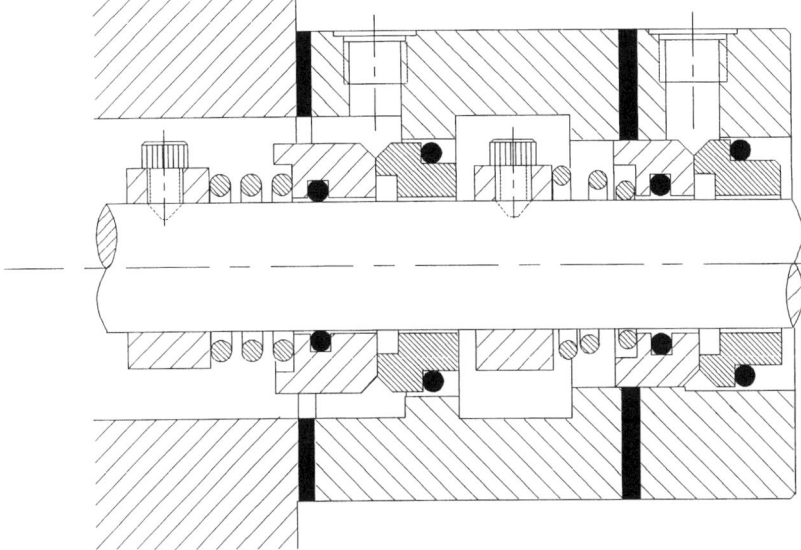

Figure 10.8 Tandem Mechanical Seal

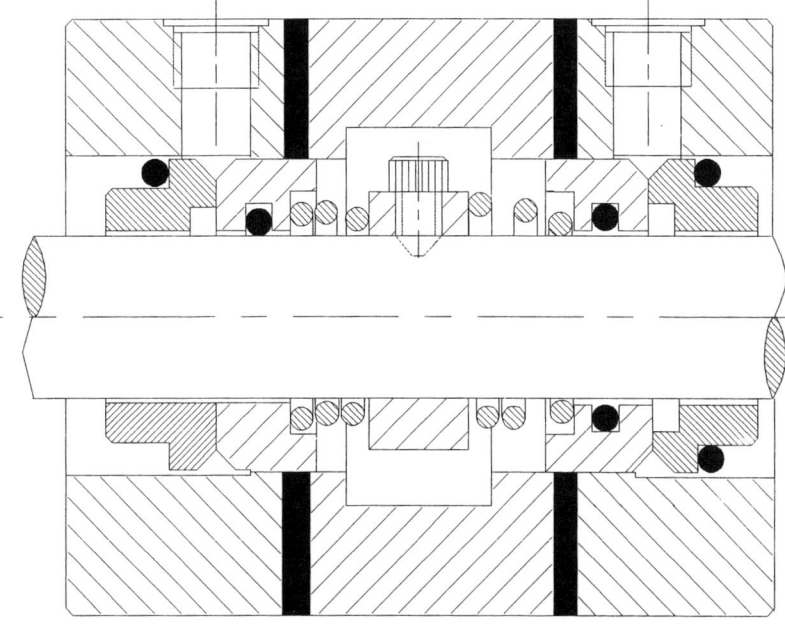

Figure 10.9 Double Mechanical Seal

- Use BDPFX if neither dry crude nor clean fresh water is available. X in the seal next to the product shall be tungsten carbide against tungsten carbide (TC vs. TC). X in the outer seal shall be carbon against silicon carbide (C vs. SC).

Cyclone Separator

When flushing a mechanical seal with pumped fluid that contains abrasives or other contaminants, a hydrocyclone separator installed in the flush line shall clean the fluid. Do not use hydrocyclones when the pressure differential in the unit is less than 25 psi.

Flush and Quench Fluids

You may choose to arrange flushing and quenching piping according to *API Standard 610*.

Stuffing-Box Cooling

Cooling stuffing boxes is essential for pumps handling liquids at 212°F or higher. To do this, either run water through a jacket around the bearing housing or recirculate the flushing fluid through a heat exchanger.

Buffer Fluid Schemes

Tandem and double mechanical seals need buffer fluid, which may be any clean fluid, preferably water. The design pressure for the vessel holding a tandem buffer fluid shall be strong enough to withstand maximum stuffing-box pressure. The system for a double mechanical seal buffer fluid shall be able to withstand at least 50 psig more than the pump's suction pressure.

Face Seal Life Expectancy

Literature by seal manufacturers claim mechanical seals operating on design conditions may last up to 100,000 hr. This equates with at least ten years' service, which may be overly optimistic. Seals rarely run for ten years under design conditions. You can rarely avoid changes in process, liquid characteristic, or other factors. A more realistic time frame may be from two to five years.

Chapter 11
Pump Bearings

Pumps have bearings that permit shaft rotation without undue friction, which could damage the shaft, and prevent radial and axial movements of the shaft. These bearings may be internal, that is, placed inside the pump casing and product lubricated. You'll find external bearings in independent bearing housings fastened to the pump's casing. Either oil or grease lubricates external bearings.

All horizontal pumps have at least two bearings, called the inboard bearing and the outboard bearing. The inboard bearing, located next to the drive coupling, usually functions as a radial bearing. All centrifugal and rotary pumps also have a thrust bearing, which usually is the outboard, even though this bearing may also act as a radial bearing and may be a single-row antifriction bearing exactly like the outboard.

Pumps of different sizes and services have a variety of bearings, and some are mentioned below.

Bearing Types

Sleeve or Journal Bearings

Even though many small pumps have radial sleeve bearings, antifriction bearings are preferred in small- and medium-sized pumps. Many companies have rules of thumb for allowing the use of radial ball bearings. For example, antifriction bearings are allowed when the product of the rated pump kilowatt and the pump rpm does not exceed 1.2×10^6 for between-bearing pumps and 0.9×10^6 for overhung impeller pumps. Large, high-speed pumps have oil-lubricated babbitt sleeve bearings at both the outboard and inboard ends of the shaft. The outboard end also has a tilted-pad thrust bearing. When these journal bearings start and stop, the oil film between the bearings may be cut, causing increased friction and wear. Therefore, these journal bearings are usually pressure lubricated. One rule of thumb is to pressure lubricate the bearing when the shaft peripheral speed exceeds 45 ft/sec. (See Figure 11.1.)

Internal Product-Lubricated Bearings

These internal bearings frequently appear in small, inexpensive pumps in clean liquid service. These are inexpensive sleeve bearings, more like bushings than anything else. A French multistaged, high-pressure pump also has internal hydropneumatic bearings.

84 *Practical Introduction to Pumping Technology*

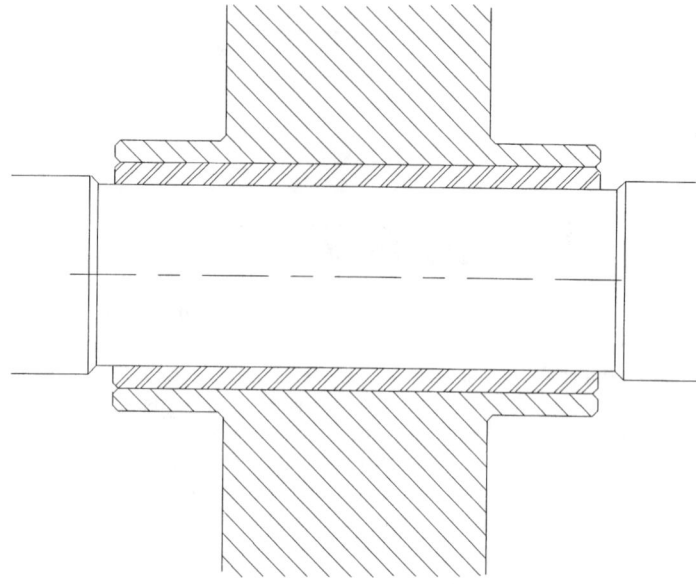

Figure 11.1 Internal Sleeve Bearing

Antifriction Bearings

Single-Row Antifriction (Ball) Bearings

Pump manufacturers should design all antifriction bearings (Figure 11.2) for 100 percent maximum load. The bearings shall have a specified B-10 life. ANSI pumps usually specify three years or 25,000 hr. *API Standard 610* pump bearings have a theoretical B-10 life of 45,000 hr.

Only a small percentage of pump bearings last their B-10 period. Most bearings fail before then because of improper lubrication, poor bearing housing sealing arrangements, and faulty installations. Because bearings demand a tight fit, even a slightly tilted bearing will overheat and fail.

Single-row, deep-grooved ball bearings are common in medium and small overhung pumps and axially split pumps, both as radial and thrust bearings. ANSI and API pumps use them extensively as radial bearings.

Double-Row Antifriction Bearings

These two rows of deep-grooved antifriction bearings (Figure 11.3) are contained within the same bearing race. Their application may be as a radial or a thrust bearing. The latter is more common, as seen, for instance, in many API pumps.

Self-aligned Double-Row Antifriction Bearings

Self-aligning bearings have an application in high-speed pumps with slightly flexible shafts. These bearings do not handle thrust well, which limits their use as radial bearings.

Pump Bearings **85**

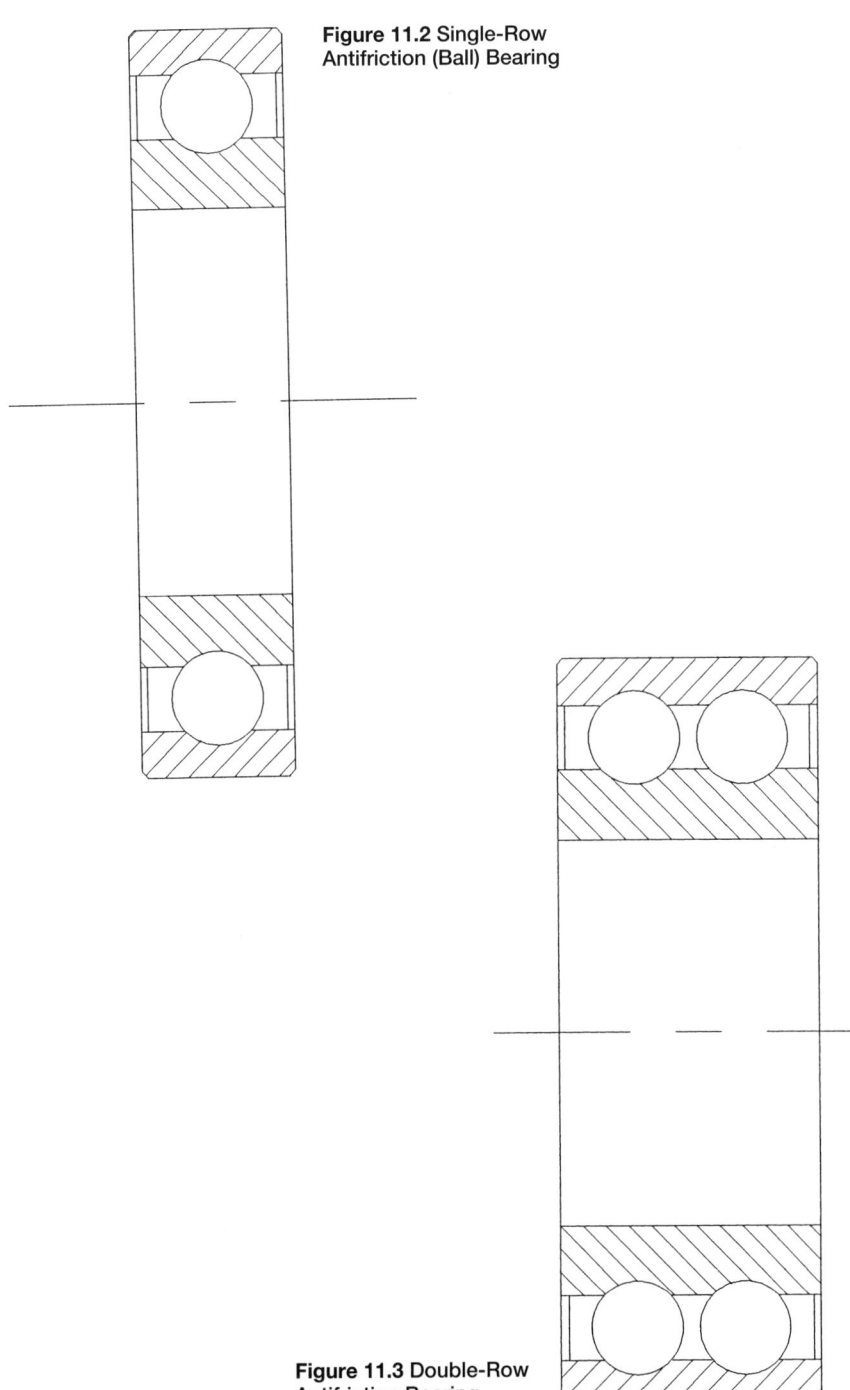

Figure 11.2 Single-Row Antifriction (Ball) Bearing

Figure 11.3 Double-Row Antifriction Bearing

Angular Contact Bearings

The design of 40° angular contact bearings makes them suitable for thrust bearings. A single-row angular thrust bearing (Figure 11.4) only accepts thrust from one direction. In most centrifugal pumps, the thrust reverses during start-up, so the thrust bearing must absorb thrust from both directions. Double angular contact bearings placed back to back within the same outer race will do that (Figure 11.5). A better solution for a pump thrust bearing may be to use two single, slightly preloaded angular contact bearings placed back to back.

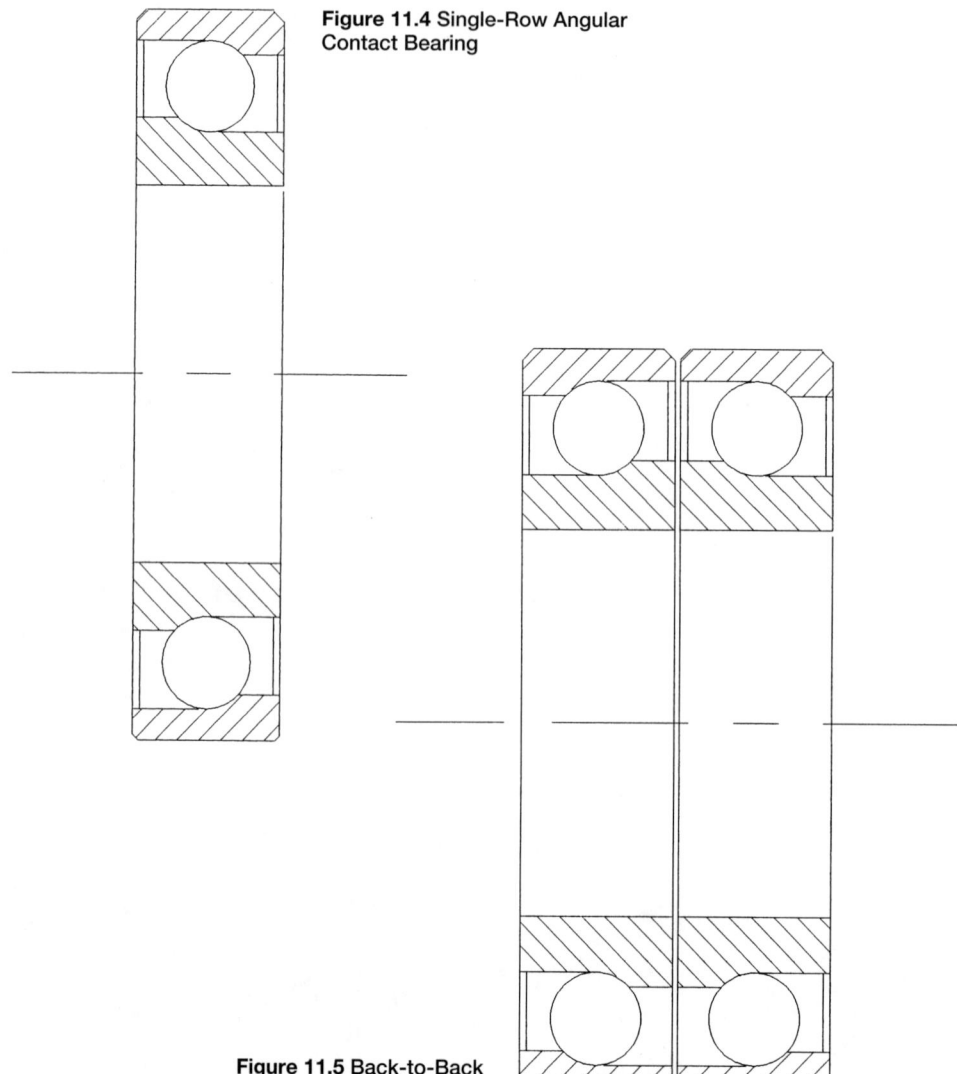

Figure 11.4 Single-Row Angular Contact Bearing

Figure 11.5 Back-to-Back Angular Contact Bearing

Roller Bearings

Many companies do not permit the use of roller bearings (Figure 11.6) in pumps. In pumps with very large shaft diameter, their use may be accepted only as radial bearings. Roller bearings do not accept thrust well.

Tilted-Pad Bearings

This type of bearing (Figure 11.7) consists of tiltable pads or segments that rest on hard steel anchors in the bearing housing. The pads are constructed so they can rock slightly backward and forward. This movement helps the pads form a wedge-shaped oil film between the pads and the shaft, which breaks the thrust load on the shaft. Tilted-pad bearings are usually thrust bearings, but some companies recommend their use as radial bearings in special cases. For instance, one company mandates their use on pumps with rated speeds of more than 180 percent of the first lateral critical speed.

A Kingsbury thrust bearing is a double-acting, multiple segment tilted-pad bearing. The center-pivoting pads can absorb thrust in both directions. Mitchell manufactures a similar bearing in Europe.

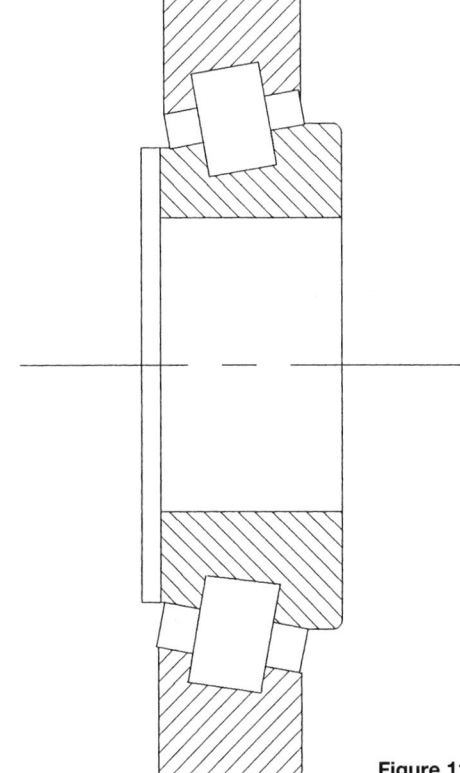

Figure 11.6 Roller Bearing

88 Pumping Primer

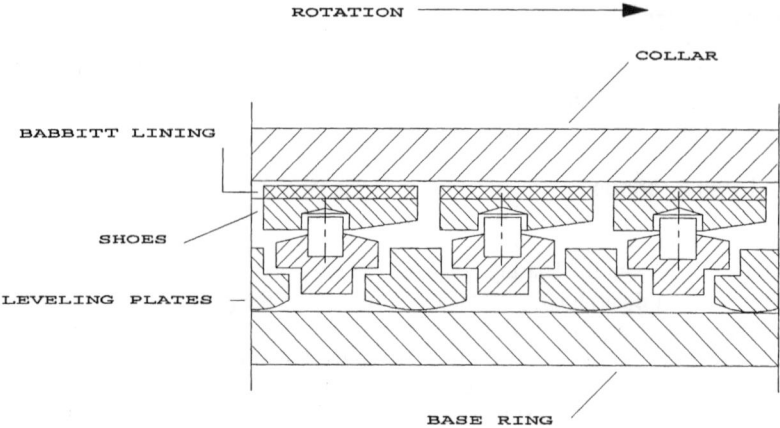

Figure 11.7 Tilted-Pad Thrust Bearing

Vertical Pump Bearings

Thrust Bearings

Both vertical turbine and vertical in-line pumps may have their thrust bearings either in the driver or in a separate pump bearing housing. If the thrust bearing is in the driver (electrical motor or right-angle drive), the bottom end of the pump shaft turns in a product-lubricated bronze or carbon bushing. (See Figure 11.8.)

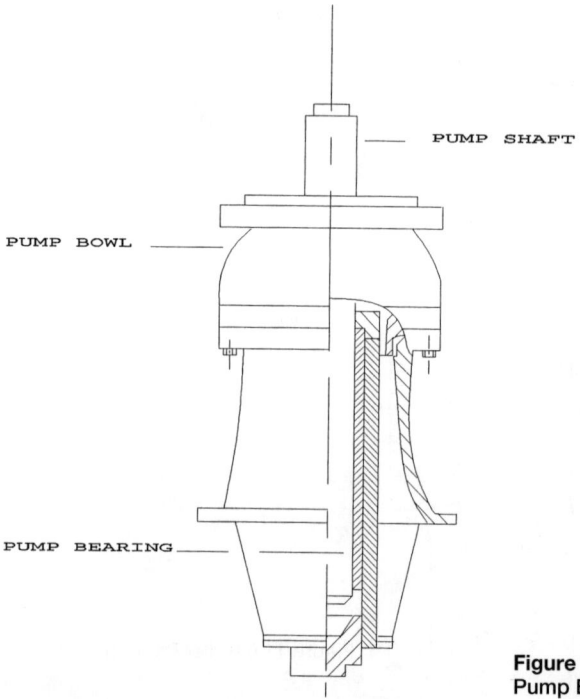

Figure 11.8 Vertical Turbine Pump Bowl Bearing

Line-Shaft Bearings

Sleeve bearings spaced at regular intervals and held in place by retainers help keep whip, or axial movement, out of the line shaft on vertical turbine pumps. A common practice is to place one bearing each 5 ft from the top and bottom of the shaft. A spacing of 10 ft from additional bearings will suffice. These bearings are either product lubricated in an open-shaft construction or oil lubricated if the shaft is enclosed. If these bearings are cut rubber, you cannot use a carbon steel shaft. To avoid corrosion, a monel or stainless steel line shaft is necessary. A wise person will prelubricate the bearings in an open-shaft pump to avoid premature wear of these. (See Figure 11.9.)

Bearing Lubrication

Grease Lubrication

On most small- and medium-sized pumps, grease lubricates the antifriction bearings. Most of these bearings are sealed, prelubricated, and maintenance free. The

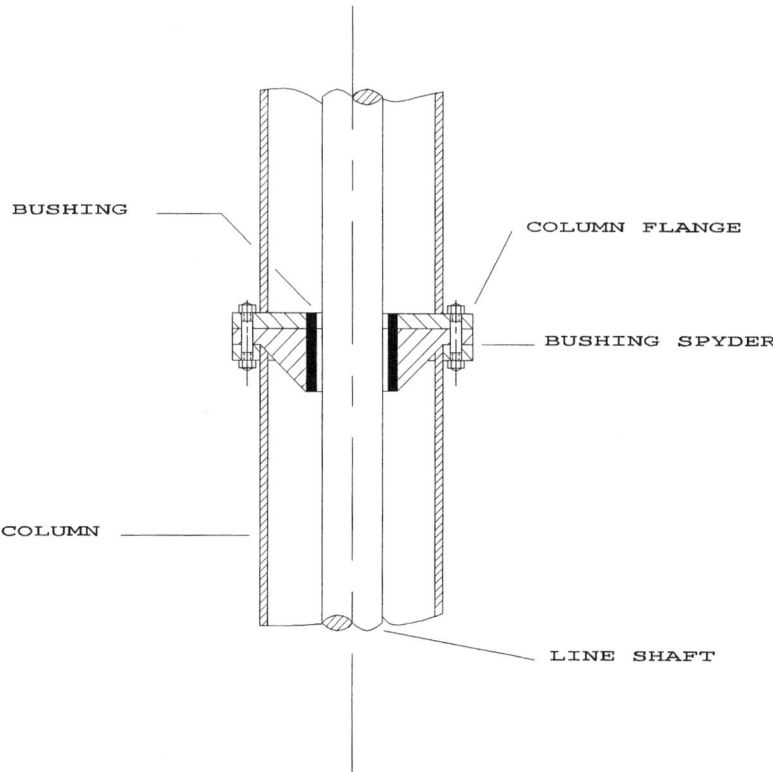

Figure 11.9 Vertical Line-Shaft Pump Bearing Arrangement

cold grease occupies about a third of the space in the bearing. As the bearing turns and heats up, the grease turns to liquid. The outer race acts as a heat exchanger and keeps the grease temperature within acceptable limits.

Oil Lubrication

Oil-lubricated ball bearings may be:

- Ring oil lubricated (Figure 11.10), with two or more oil rings guided by machined grooves on the shaft. Throwers deposit a fine oil mist on all the bearing surfaces. Seals keep out dirt and other contamination. A constant-level lubricator keeps the oil level consistent.
- Splash lubricated
- Mist lubricated
- Forced-feed lubricated, where pressurized oil from an outside source circulates through the bearing housing. In the simplest system, the oil pumps through a filter and a heat exchanger into an oil reservoir. In more complex installations, the lube oil design may be according to *API Standard 614* or a variation thereof.

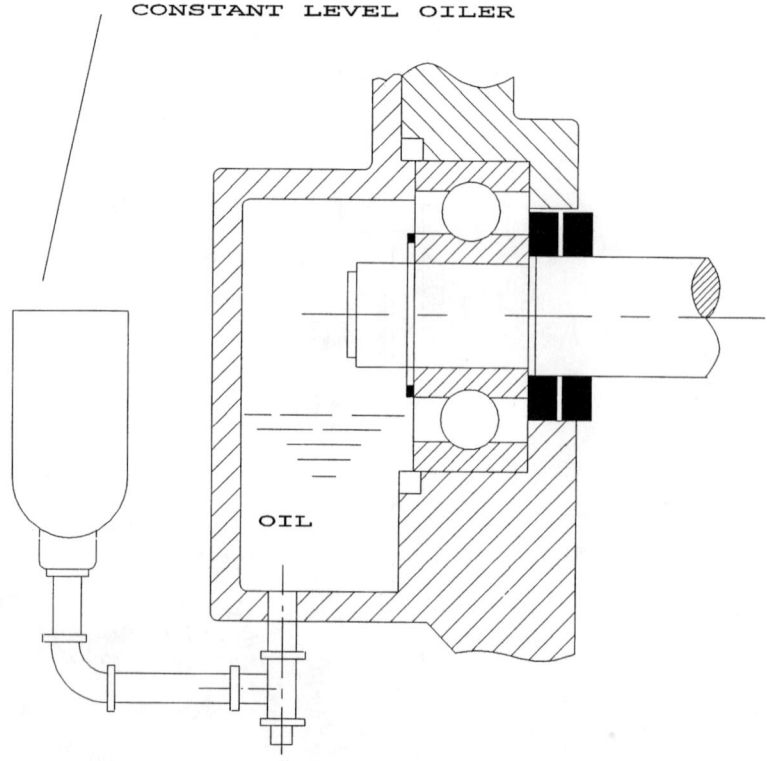

Figure 11.10 Ring Oil Lubrication With Constant-Level Oiler

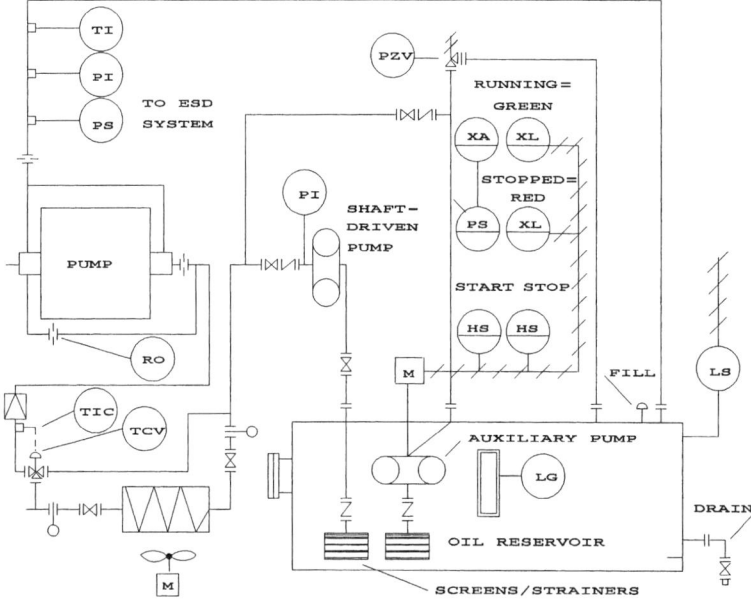

Figure 11.11 *API Standard 610* Type Lubrication System

Forced-feed lubrications shall have provisions for supply in case of shutdown or failure of the main lube oil pump. This may be an auxiliary lube oil pump that also may act as a prelube pump, accumulator, or overhead tank. (See Figure 11.11.)

Bearing Cooling

API Standard 610 requires that bearing oil temperatures shall be below 180°F for ring oil or splash systems, based on specified operating conditions and 110°F ambient temperature. In pressurized systems, the oil temperature shall not exceed 160°F. Packings shall be cooled when the fluid temperature exceeds 300°F.

The company's piping specifications determine the materials in the cooling piping. Pumps in critical services may warrant flow indicators in their cooling piping.

Bearing Seals

Lip seals and labyrinth seals keep dirt and other contaminants out of the bearing housing. Shaft rotation against the elastomer lips rapidly wears down this type of seal. A reliable substitute is the so-called magnetic seal, which consists of a stationary steel ring with an O-ring seal attached to the bearing housing and a ferromagnetic stainless steel floating ring. Sometimes manufacturers add an expansion chamber to this arrangement to accommodate varying internal pressures in the sealed bearing housing.

Chapter 12
Metallurgy

With the enormous varieties of fluids that pumps move at ever-increasing pressures, choosing the right pump metallurgy has become critical. In ancient Egypt, Archimedes screw pumps were made out of wood, and most still are. The first centrifugal pumps mostly handled water at low or moderate pressures. These pumps had an all-iron construction, with a carbon steel shaft. Small water pumps in sweet, clean water services still widely use this construction.

Later, copper alloys were introduced, often in pumps with cast iron casing and bronze impellers. Copper alloys, such as monel and nickel aluminum bronze, are modern metals used in aggressive services. All pump manufacturers and industry standards have tables showing the metallurgy recommended for different services. One of the most complete is the one published in *API Standard 610, Centrifugal Pumps for General Refinery Services*.

Corrosion

Apart from cavitation and possible erosion, corrosion is the major consideration when selecting pump material. Corrosion problems occur mostly in pumps that use fluids with high sodium chloride content. The liquid, more aggressive if acidic, contains dissolved oxygen and/or other impurities, such as hydrogen sulfide (H_2S) and ammonia.

Metals exposed to oxygen form a covering layer, also called a passive layer, of oxide, which protects the metal from corrosion. The various types of stainless steel make up the largest group of metals capable of forming thick passive layers. Types of metal corrosion are:

- General uniform corrosion
- Galvanic corrosion
- Pitting
- Crevice corrosion
- Erosion corrosion
- Cavitation corrosion
- Stress corrosion
- Vibration crack corrosion

You can easily prevent general uniform corrosion by using the correct metallurgy, such as stainless steel or nickel alloy steels. Materials in contact with or relatively close to each other that have different electrochemical potentials cause galvanic corrosion. The material with the lower potential will ultimately dissolve, especially in fluids with high electrical conductivity. As a solution, avoid dissimilar metals. When this is unavoidable or not economical, use an isolation kit. This is commonly seen with seawater intake pumps on offshore production platforms, where the vertical turbine pump may be stainless or duplex steel, and the column and discharge head are carbon steel.

Changes in the fluid's chemical composition or defects in the passive film may cause local corrosion, such as pitting and crevice corrosion. Local corrosion may occur in standby pumps in aggressive services. To maintain the passive film, the metal needs a constant supply of oxygen. In standby pumps partially filled with stagnant liquid, this does not occur. Pitting and crevice corrosion results. As a solution, purge and flood these pumps with fresh water when on standby.

In modern pumps, the velocities of the fluids and the enclosing metals are high. This may cause mechanical erosion if the fluid is dirty or corrosion if high, localized velocities disturb the passive film. To remedy this, select erosion-resistant material. For instance, a type 316 stainless steel impeller is more resistant than a Ni-resist construction.

Loss of pump material due to cavitation, the implosion of bubbles in the liquid, occurs in all pumps, but the consequences are more serious in an aggressive service. Because the implosion destroys the metal's protective cover, not only does mechanical erosion damage the pump, but aggressive local corrosion takes place and accelerates the pump's destruction. Ductile materials that also have high strength combined with hardness—regardless of the chemical composition—will better resist cavitation damage.

Stress corrosion and vibration crack corrosion are not usually seen in pumps. However, if the fluid exceeds 150°F, stress corrosion is possible. In these instances, pump designers must guard against stress corrosion by avoiding sharp transitions and selecting materials resistant to pitting.

Table 12.1 shows some of the most common materials used in the pump industry.

Pump Materials

Cast Iron

The sheer quantity of pumps made from some type of cast iron that are used in low-performance water services worldwide makes cast iron the most widely used pump material. Iron casts easily, costs relatively little, and resists the elements in outdoor use. Its low tensile strength makes it unsuitable for high pressures and erosion caused by abrasives in the media. Bronze or low-alloy carbon steel impellers in a cast iron casing will increase pump impeller life. Cast iron pumps have carbon steel shafts.

Ferritic Steel

Carbon steel is iron with a small amount of carbon added. The steel requires full ferritizing heat treatment. All castings shall also have stress-relief heat treatment, according to their respective ASTM standards. Castings made of ASTM-A216

(text continued on p. 97)

Table 12.1
Pump Materials

Material (ASTM)	Castings	Forgings	Rods, Bars
Cast Iron			
Gray cast iron, nonpressurized	A48 Cl 30 or 40		
Gray cast iron, pressurized	A278 Cl 30 or 40		
Ductile iron, nonpressurized	A536		
Ductile iron, pressurized	A395		
Austenitic cast iron (Ni-resist)	A436		
Ferritic Steel			
Carbon steel, 32°F	A216 WCB or WCC	A105	A106, A672
Carbon steel, −45°F	A352 Gr LCC	A516, A537	A322 Gr 4130, 4140
Nickel steel, 2½	A352 Gr LC2	A350	A322 Gr 4340
Nickel steel, 3½	A352 Gr LC3		
Martensitic Stainless Steel			
Tp 410–12% Cr	A743 CA-15	A182	A276 410
12% Cr–4% Ni	A487 CA-6NM	A182	
17-4 pH–17Cr, 4Ni	A743 CD-4MCu	A705	A564
Austenitic Stainless Steel			
Tp 304–19Cr, 9Ni	A743 CF-8	A184 F04	A276 304
	A744 CF-8	A473 304	
Tp 304L–18Cr, 8Ni (low C)	A743 CF-3	A184 F304L	A276, 304L
	A744 CF-3	A473 304L	
Tp 316–16Cr, 12Ni, Mo	A743 CF-8M	A182 F316	
	A744 CF-8M	A473 304L	
Incoloy 800–21Cr, 32Ni	A351 HT-30	B564	B408
Alloy 20–20Cr, 29Ni, Mo, Cu	A743 CN-7M	B462	B473
	A744 CN-7M	B472	
254 SMO–20Cr, 18Ni, Mo, Cu		A182 F44	A276
Nitronic 50–22Cr, 13Ni, 5Mn, N	A351 CG6MMN	A182 FXM 19	A276
	A743 CG6MMM		A479
Nickel Alloy			
Monel 400	A494 M 35 1	B564	B164
	A743 M 35 1		
Inconel 600–15Cr, 8Fe, balance Ni	A494	B564	B166
	A743		
Hastelloy B-2–28 Mo, balance Ni	A494 N12MV	B366	B335
Hastelloy C-276–16Mo, 15.5 Cr, 5Fe, 4W, balance Ni	A494 CW12MW	B366	B574
	A473 CW12M	B574	
Copper Alloy			
70-30 copper nickel	B369 C96400		B122
90-10 copper nickel	B369 C96200		B122
7% aluminum bronze	B148 C95600	B150	B150
9% aluminum bronze	B148 C95200	B283	B124
5% nickel aluminum bronze	B140 C95800	B124	B124
Zincless bronze–9Sn, 2Pb	B505 C92700		

Table 12.2
Duplex Stainless Steel Compositions

		Cr	N	Mo	Cu	N	Other
S32404	Uranus 50	21	7	2.5	1.5		
J83370	CD-4MCu	25	5	2.0	3.0		
S32550	Ferralium 225	25	6	3.0	2.0	2.00	
J93404	Atlas 958	25	7	4.5		0.25	
S31260	DP-3	25	7	3.0	0.5	0.15	0.3W

Table 12.3
Recommended Material for Corrosive Liquids

Liquid	Excellent	Good	Not Suitable
Acetic acid, 70°F	Bronze, 316 SS	Monel	Cast iron
Amines and glycols	316L SS	316 SS	CI, CS
Ammonium chloride, <10%, 70°F	Titanium	Hastelloy C	CS, CI
Arsenic acid, to 200°F	Carpenter SS	316 SS	CS, Cu
Boiler feed water	13% Cr	CS	CI
Boric acid, to boiling	Carpenter SS	316 SS	CS, CI
Brine, acidic, 70°F	Hastelloy C	Duplex SS	CS, CI
Bromine, dry, 70°F	Monel	Hastelloy B	316 SS, CS
Calcium chloride, 70°F	Hastelloy C	Hastelloy B	Bronze
Carbonic acid, 212°F	Hastelloy C	316 SS	
Carbon tetrachloride	Monel	316 SS	
Chloric acid, 70°F	Carpenter SS	CD4MCu	Other
Chlorinated water	Titanium	Hastelloy C	
Citric acid	Hastelloy B	316 SS	CS, CI
Crude oil, dry	13% Cr	CS	CI
Crude oil, wet > 0.2% H$_2$O, sweet and sour	CS	316 SS	CI
Cupric chloride	Titanium	Hastelloy C	316 SS, CS
Diesel oil, No. 2			
Diethylene glycol, 70°F	316 SS, CI	Hastelloy C	
Ether, 70°F	316 SS	CI	
Ethyl alcohol, to boiling	316 SS	CI	
Ethylene glycol	CI, CS	316 SS	
Ethylene oxide, 70°F	316 SS	Monel	Bronze
Ferric chloride, <5%, 70° F	Titanium	Hastelloy C	All other
Ferric nitrate, 70°F	Hastelloy C	316 SS	Monel, CS
Fire water, salt	Duplex SS	Al Bronze	CI, CS
Fire water, sweet	CS	Bronze	CI
Formaldehyde, to boil	316 SS	CI, CS	
Formic acid, 212°F	Carpenter 20	CD4MCu	316 SS
Fresh drinking water	CS	CI	
Hydrochloric acid, <1%, 70°F	Hastelloy C	Hastelloy B	CI, CS, Bronze

(table continued on next page)

(*table continued from previous page*)

Table 12.3 (continued)
Recommended Material for Corrosive Liquids

Liquid	Excellent	Good	Not Suitable
Hydrogen peroxide, 150°F	316 SS	Monel	Bronze
Lactic acid, 50%, 70°F	316 SS	Bronze	CI, CS
Lime slurry	CD4MCu	CI, CS	
Magnesium chloride, 70°F	Carpenter 20	316 SS	
Magnesium hydroxide	Bronze, 316 SS	CI, CS	
Magnesium sulfate	316 SS	Monel	
Methyl alcohol	Any	Any	
Nitric acid, 70°F	316 SS	Titanium	CI, CS, Bronze
Nitrous acid, 70°F	Carpenter 20	316 SS	
Nitrous oxide	Carpenter 20	316 SS	Monel, Ni
Oleic acid	Carpenter 20	Hastelloy C	
Oleum, 70°F	316 SS	CI, CS	
Oxalic acid	Carpenter 20	316 SS	CI, CS
Phosgene, 70°	Carpenter 20	316 SS	
Phosphoric acid, <19%, 70°F	Carpenter 20	316 SS	CI, CS
Piric acid	Carpenter 20	316 SS	CI, CS, Bronze
Potassium carbonate	Bronze, 316 SS	Monel	CI
Potassium chloride, 70°F	Carpenter 20	Nickel	
Potassium cyanide, 70°F	316 SS	Monel	
Potassium hypochlorite	Hastelloy C	Carpenter 20	CI, CS
Potassium phosphate		316 SS	
Salt water (sea)	Duplex SS	Carpenter 20	
Sodium bisulfite, 70°F	Carpenter 20	316 SS	CI, CS
Sodium bromide, 70°F		316 SS	
Sodium chloride, 70°F	Carpenter 20	316 SS, Bronze	
Sodium cyanide	316 SS	CS	Br, Ni, Monel
Sodium hydroxide, 70°F	316 SS	CI, CS	
Sodium hypochlorite		Carpenter 20	CI, CS, Bronze, Monel, Ni
Sulfur, molten	316 SS	CS	
Sulfuric acid, >10%, 175°F	Carpenter 20	CD4MCu	CI, CS, 316 SS, Ni, Monel, Titan
Sulfuric acid, >3–4%, to boil	Carpenter 20	Hastelloy B	CI, CS, Bronze, 316 SS, Monel, Ni, Titanium
Sulfuric acid, 70°F	316 SS	Hastelloy B	CI, CS, Monel, Ni
Titanium tetrachloride	Hastelloy B	Carpenter 20	
Trichlorethylene	Titanium	316 SS	
Urea, 70°F	Carpenter 20	316 SS	
Vinyl chloride	Titanium	CS	
Zinc cyanide, 70°F	Carpenter 20	Monel	CI, CS
Zinc sulfate	Carpenter 20	316 SS	CI, CS

(*text continued from page 93*)

Grade WCB carbon steel suit fluids above 32°F. ASTM-A352 Grade ICC3 will withstand temperatures as low as −50°F. Carbon steel suits high pressures and a wide variety of fluids; it is hard and resists abrasion but is not adequate for high chloride service. Carbon steel is sometimes used in seawater service but only with de-aerated water. A popular carbon steel shaft material is ASTM A322, Grade 4130 or 4140.

Only the addition of nickel makes ferritic steels resistant to local corrosion. Such steels used in pump construction include the 2½ (−100°F) and 3½ (−150°F) nickel steels. All ferritic steels should be Charpy V-notch impact tested at their designated temperatures.

Martensitic Stainless Steel

When choosing a corrosion-resistant stainless steel, analyze the following in the pumped liquid:

- Chloride content
- pH
- Oxygen content
- Contaminants

One advantage of martensitic stainless steel is its high resistance to cavitation and cavitation corrosion. It has fair resistance to local corrosion in media with chloride presence. The type 400 stainless steels are martensitic steels. In cold applications demanding stainless steel, a good choice is 17-4 pH (ASTM A743-CD-4MCu for castings). This martensitic steel, works well for up to −150°F in heat-treated conditions. Do not use martensitic steels in lethal services.

Austenitic Stainless Steel

The choice of material in modern high-performance pumps is important because of the high fluid velocities in the pumps. In corrosive services, the choice becomes critical. Where exotic materials are ruled out, only high austenitic steels, austenitic-ferritic steels, and some nonferrous materials are adequate. Austenitic cast irons, such as Ni-resist, may be used only on low-velocity pumps.

The use of austenitic steels, such as type 304 stainless steel (18Cr, 8Ni) or type 316 stainless steel (17Cr, 112Ni, Mo), is common in the food, pharmaceutical, and petrochemical industries because of its resistance to corrosion and acidity.

For more severe services, manufacturers use cast materials, such as:

- Alloy 20   (20Cr, 29Ni, Mo, Cu)
- Incoloy 825   (20Cr, 42Ni, Mo, Cu)
- Avesta SMO 254   (20Cr, 18Ni, Mo, N)
- Nitronic 50   Much in demand for shaft material
- CN7M

In austenitic stainless steel, the combination of chrome and molybdenum gives local corrosion resistance. Molybdenum-free alloys perform poorly in chloride media. Only adding 3 percent to 5 percent molybdenum a high chrome (20 percent) steel will improve the corrosion resistance significantly.

Austenitic stainless steels are not particularly resistant to cavitation and cavitation corrosion.

Duplex Stainless Steel

In the early 1980s, various steel companies developed a steel with both a ferritic and austenitic molecular structure, called duplex stainless steel. It combines the characteristics of ferritic steels, such as high yield strength and good weldability, with resistance to general corrosion, pitting, and stress-corrosion cracking. Duplex steels from different mills each have slightly different chemical compositions (Table 12.2). A typical composition for shaft material may be (Sandvik SAF 2205):

- C (max.) = 0.03%
- Si (max.) = 1%
- Mn (max.)= 2%
- P (max.) = 0.03%
- S (max.) = 0.02%
- Cr   = 22%
- Ni   = 5.5%
- Mo   = 3%

Pitting and crevice corrosion protection depends on the chromium and molybdenum present in the steel. Because of its ferritic properties, duplex steel has a high resistance to erosion corrosion. The relatively low nickel content may make duplex steel more economical than austenitic steels. That, of course, depends on the pricing policies of the steel mills. Duplex stainless steels resist cavitation and cavitation corrosion.

Nonferrous Materials

Other materials used in highly corrosive and acid environments include the following (see also Table 12.3):

Nickel-based alloys, such as:

- Monel alloy 400
- Monel alloy K500
- Inconel 600 (15Cr, 8Fe, balance Ni)
- Inconel 625 (21.5Cr, 3Fe, 9Mo, 4Nb, balance Ni)
- Hastelloy

Copper-based alloys, such as:

- 70-30 copper nickel
- 90-10 copper nickel

- 7 percent aluminum bronze
- 9% aluminum bronze
- 11 percent aluminum bronze
- Nickel aluminum bronze (5% Ni)
- Zincless bronze (9Sn, 2Pb)
- Navy bronze (6Sn, 5Pb, 4.5Zn)

Copper alloys resist biofouling and are often used as material in water intake pumps. Biofouling, such as barnacle and other shellfish, is a serious problem in salt water installations. Even with hypochlorite injection at the pump intake, it is difficult to eradicate the problem. The 90-10 copper nickel alloy has shown a remarkable resistance to this problem. The aluminum bronzes have high cavitation resistance.

Titanium

This has high resistance to all forms of corrosion and also has high cavitation resistance. Using improper gasket material may cause chloride stress corrosion even in titanium pumps.

Plastic

The use of plastic material (Table 12.4) in the chemical and food industry is widespread. In these pumps, all wetted materials are plastic, but shafts and bearings remain metallic. As a main advantage, the plastic material can resist both corrosion and aggressive chemicals. As a disadvantage, they can neither withstand high pressures nor high liquid velocities. Abrasion resistance is low; limit liquid temperatures to approximately 260°F.

Table 12.4
Plastic Pump Materials

Material	Description	Max. Temperature °F	Not Suitable For
PVC	Polyvinylchloride	140	Ketones, esters, aromatic hydrocarbons
CPVC	Chlorinated polyvinylchloride	230	Ketones, esters, aromatic hydrocarbons
PE	Polyethylene	200	Oxidizing acids
PP	Polypropylene	260	Oxidizing acids, chromic acid
PTFE	Polytetrafluoroethylene	500	
GRP	Glass-reinforced vinyl ester	250	

Elastomer Linings

Some pumps in aggressive chemical service are elastomer-lined metal pumps. The electroplating industry widely uses these types of pumps.

Chapter 13

Pump Drivers

The ancient civilizations used men or animals to power the Archimedes screw pumps they used to bring water from the irrigation ditches to their fields. Today, machinery powers most pumps. These machines include:

- Electric motors
- Internal combustion engines
- Gas turbines
- Steam turbines
- Hydraulic drives
- Solar power

Electric Motors

Most motors that drive pumps are alternating current, single or three phase, either induction or synchronous. Most suit full-voltage, across-the-line starting.

Voltage

Most electric motor drivers are alternate-current (A-C), squirrel-cage induction motors. Solar-powered pump motors are usually direct-current (D-C). Some large electric motors are synchronous. The system voltage may vary from 120 volts (v) to 13,800 volts. Table 13.1 shows suggested voltage and power ratings for induction motors.

Table 13.1
Suggested Voltage and Horsepower Ratings for A-C Motors

Nominal System Voltage	Nameplate Motor Voltage	Number of Phases	Horsepower Range	Motor Type
120	115	1	–½	Split phase
240	230	1	–¼	Split phase
240	230	3	¼–50	Squirrel cage
480	460	3	¼–250	Squirrel cage
2,400	2,300	3	250–1,500	Squirrel cage
4,160	4,000	3	500–5,000	Squirrel cage
6,900	6,600	3	500–6,000	Squirrel cage
13,800	13,200	3	1,400 plus	Squirrel cage
13,800	13,200	3	10,000 plus	Synchronous

In the United States, the standard A-C wavelengths are 60 cycles. Overseas, many distribution systems use 50 Hz. In 50-Hz systems, both motor speeds and voltages differ from 60-Hz systems. Table 13.2 shows speed and voltages for two-pole to six-pole 50-Hz and 60-Hz electric motors.

Table 13.2
Full-Load Speeds of A-C Motors

Horsepower	Number of Poles	Supply Frequency (Hz) 50	60
1–3	2	2,820	3,380
	4	1,405	1,685
	6	930	1,115
4–10	2	2,855	3,425
	4	1,420	1,700
	6	945	1,130
15–30	2	2,910	3,490
	4	1,445	1,730
	6	960	1,145
40–100	2	2,935	3,520
	4	1,475	1,770
	6	975	1,170

Power

One motor horsepower is the shaft output of one American Standard horsepower unit, equal to 746 watts (w).

Applicable Codes

- *ANSI/NEMA MG1 Motors and Generators*
- *ANSI/UL 674 Electric Motors and Generators for Use in Hazardous Areas*
- *ANSI/NFPA 70 National Electric Code*
- *IEC 24 Rotating Electrical Machines*
- *IEC 72 Dimensions and Output Ratings for Rotating Electrical Machines*
- *IEEE 112 Test Procedures for Polyphase Induction Motors and Generators*
- *IEEE 114 Test Procedures for Single-Phase Induction Motors*
- *IEEE 115 Test Procedures for Synchronous Machines*
- *IEEE 421 Criteria and Definitions for Excitation Control Systems for Synchronous Machines*
- *ANSI C50.10 General Requirements for Synchronous Machines*

Hazardous Locations

An area with a large concentration of flammable gases, vapors, dust, or fibers that may cause an explosion or fire is a hazardous location. Class, division, and group are classify hazardous locations. The *National Electrical Code, Article 500*'s definition of a classified area is paraphrased below.

Class I
Areas where flammable gases or vapors may cause an explosion or a fire. Some Class I locations are found in:

- Petroleum refineries.
- Oil and gas production facilities.
- Liquified petroleum gas (LPG) and liquified natural petroleum gas (LNPG) plants and storage facilities.
- Petrochemical plants.
- Dry cleaners.
- Aircraft fueling stations and hangars.

Class II
Areas with combustible and flammable dust. Some Class II locations are found in:

- Grain mills and storage facilities.
- Plastic plants.
- Sugar plants.
- Paint manufacturing plants.
- Paint spray booths.

Class III
Areas with flammable fibers in the air, or collected on floors and machinery. Class III areas include:

- Saw mills.
- Wall board plants.
- Masonite plants.
- Chip board plants.

Division 1
An area where during normal conditions, such as regular operations and/or maintenance, the hazards are present.

Division 2
An area where the hazards exist only during abnormal conditions, such as tank leakage or upset process.

Groups A, B, C, D
Classifications according to the flammability and explosion capability of the hazardous material in a classified area. Group A is the least flammable.

Groups E, F, G

Classifications according to ignition temperature and conductivity of the hazardous material in a classified area. Group G has the highest temperature and conductivity.

Phases

Electric motors may be one, two, or three phased. Most fractional horsepower motors are single phased. Larger motors, 1 hp and up, are mostly three phased.

Motor Enclosures

The choice of motor enclosures depends on the location of the motor. Enclosures for integral horsepower motors are either cast steel, cast iron, cast aluminum, or fabricated steel no less than 0.125 in. thick. Treat all enclosures to withstand corrosion. Aluminum enclosures are not recommended on motors larger than 25 hp.

The type number refers to the *NEMA Standard for Industrial Enclosure*. The type locations and features of the most common motor enclosures are:

Open

This enclosure allows cooling of windings through natural convection. The enclosures suit indoor equipment in a dry, clean environment.

Open Drip-Proof

This enclosure's openings permit water or debris to fall on the motor at an angle no greater than 15°. Common use is outdoors under a shed, without complete exposure to the elements.

Weather-Protected Type I (WPI)

This open-construction enclosure features openings designed to keep out rain, snow, and sleet. The openings will not permit passage of a test rod ¾-in. outside diameter (OD) or larger. The design permits outdoor use without additional protection.

Weather-Protected Type II (WPII)

The design of the enclosure provides that the air intake and discharge will purge water and other debris blown into the enclosure by high velocity winds before they touch the windings or other electrical components. These are used on larger pumps in Class II, Division 1 areas. Offshore production and drilling platforms commonly call for use of this type of enclosure.

Totally Enclosed, Fan-Cooled (TEFC)

This motor does not have any air openings. An external fan driven by the motor shaft cools the motor exterior. An enclosure of this type is common for smaller motors in a Class II, Division 1 area. The use of WPII or TEFC motors costs relatively little. The break point between the two lies around 200 hp.

104 *Practical Introduction to Pumping Technology*

Totally Enclosed, Water-Cooled

Although similar to a TEFC, in this enclosure water jackets, cool the motor instead of a fan.

Explosion-Proof

The design has no air openings. It will withstand an internal explosion and prevent any eventual explosion from igniting flammable gases or vapors surrounding the enclosure.

Bearings

The different types of bearings commonly used in electric motors are sleeve bearings, antifriction (ball) bearings, thrust bearings, and roller bearings.

Sleeve Bearings

Fractional horsepower motors have wick oil lubricated sleeve bearings. Larger motors (larger than 1 hp) use ring oil lubricated sleeve bearings made of bronze or babbitt. All sleeve bearings shall have top and bottom half-shell bearing liners, preferably with interchangeable top and bottom liners.

Sleeve bearings are preferred on all electric motors 250 hp and above. Electric motors with a synchronous speed of 3,600 rpm and above preferably shall have sleeve bearings also.

Large motors in which the bearings generate a large heat load also have sleeve, or journal, bearings. Most of these have pressure lubrication. Often, one lubrication system is common to the motor bearings and the pump bearings.

Antifriction (Ball) Bearings

All NEMA motors have grease-lubricated ball bearings.

Thrust Bearings

All vertical motors have a thrust bearing. This may be double-row antifriction bearings, a double row of angular contact antifriction bearings, or Kingsbury thrust bearings. Both in-line pumps and vertical turbine pumps may either have the main thrust bearings in the pump, or more commonly, have the electric motor bearing absorb all the thrust generated by the pump and the motor rotors.

Roller Bearings

No application for roller bearings in electric motors exists.

Insulation

Motor windings have insulation to separate them from mechanical parts and to seal them from the exterior. The maximum allowed temperatures for NEMA insulations are:

NEMA Insulation	Temperature
A	90°C
B	130°C
F	155°C
H	180°C

All motors 1 hp and larger shall have a winding temperature rise that will not exceed 65°C for Class B insulation and 85°C for Class F at rated voltage, rated frequency, and full load. It is also prudent, in most locations, to select motors designed for 100 percent humidity.

Acceleration

The formula for pump acceleration is:

$$t = (WK^2 \times \Delta rpm) \div (T \times 308)$$

where:

t = time in seconds
WK^2 = total moment of inertia (lb-ft^2)
T = torque in lb-ft
Δrpm = speed change

Selecting motors for centrifugal pumps is fairly direct. The WK^2 of the majority of centrifugal pumps is small enough to ignore in normal applications. The torque-load curve varies with the square of the speed. Thus, at full speed, the pump torque-load equals the full torque-load of the pump.

Almost all squirrel-cage motors generate at least 70 percent starting torque. The curves rise sharply from there. Thus, with a constant power supply, the motor horsepower, rpm, and voltage only have to match the pump conditions, assuming the motors start on full voltage.

Service Factor

This term refers to the thermal capacity of the motor to operate at a higher horsepower rating than its nameplate indicates. Service factors may vary between 1 and 1.5. Most electric motors have a service factor of 1. Motors with higher service factors are useful if a pump at some interval will require more horsepower than its nominal rating. Service factors larger than 1.15 are rare.

Efficiency

The pump efficiency in percentage is shaft output power divided by electrical input power. Efficiencies for integral horsepower motors should not be lower than 93 percent.

Number of Starts

Most pumps in general pump applications will start in the following conditions:

- Two starts in sequence with initial motor temperature at 60°F
- One start with the motor temperature at its maximum temperature rating
- One additional start after the motors have cooled for 30 min

Space Heaters

Many integral horsepower electric motors have space heaters to prevent condensation in the motor casing, which may accumulate during storage or standby and could short circuit the windings. With today's epoxy encapsulated rotors, the benefit of space heaters may prove marginal. Space heaters are 120 v/1 pH/60 Hz or 220 v/1 pH/50 Hz.

Internal Combustion Engines

Because of its high efficiency, people almost always use the electric motor where electric power is available. However, a large number of pumps do exist where electric power is not available and where running electric power lines is not cost effective. There, internal combustion engines may be used. Each engine, usually the manufacturer's standard design, comes complete with heat exchanger, water pump, air and lube oil filters, lubricating pump, and governor.

Engine Types

The following internal combustion engines may drive both centrifugal and positive displacement pumps:

- Gasoline-fueled engines
- Diesel-fueled engines
- Natural-gas fueled engines

Gasoline-Fueled Engines

The use of gasoline engines to power pumps is limited to sizes unavailable in diesel. Many applications are for portable pumps in agricultural, drainage, and construction uses.

Diesel-Fueled Engines

The engines should be capable of full-load service for the required operating conditions. The engines should also be capable of operating with ASTM-grade fuel oil at the conditions specified by the project. Diesel engines may be both turbocharged and naturally aspirated, though today most engines are both turbocharged and intercooled. Most diesel engines have four cycles, but at least one large U.S. manufacturer produces two-cycle diesel engines. The largest uses for diesel-engine-driven pumps are found in:

- Fire pumps
- Irrigation pumps
- Pipeline booster pumps
- Remote water well pumps
- Oil well pumps

Natural-Gas Fueled Engines

The same criteria apply for natural-gas fueled engines as for diesel engines, differing only in the fuel supply and ignition.

Power Ratings

An internal combustion engine's basic power rating is for sea level at 60°F at rated speed with normal fuel. For any other conditions, the engine shall be de-rated according to the following information. The power ratings also depend on the service conditions for the engine. Typical engine ratings are:

Maximum

This is the highest power developed at a given speed and is the horsepower capability of the engine that can be demonstrated within 5 percent of the factory specifications under standard conditions.

Intermittent

This is the highest brake horsepower recommended by the manufacturer. The horsepower and speed capability of the engine shall permit one hour of continuous service, followed by one hour of operation at or below the continuous rating.

Continuous

This is the horsepower rating at which the engine can run continuously without interruptions or load cycling.

Many engine manufacturers use different rating schedules, such as:

- Continuous—Same as above
- Intermittent—Engines running no more than eight hours during a 24-hr period
- Standby—An engine for continuous or emergency use

Factories always rate horsepower higher for engines intended for intermittent use than for engines rated for continuous or standby use.

Engine Speed

To limit engine and piston speeds in continuous duty machines, consider these conservative approaches:

- Engines 100 brake horsepower or smaller—Maximum engine speed shall not exceed 2,600 rpm. Piston speed not shall not exceed 1,800 feet per minute (fpm).
- Engines between 100 and 600 hp—Maximum engine speed shall not exceed 2,200 rpm. Piston speed shall not exceed 1,850 fpm.
- Engines 600 hp and larger—Maximum speed for four-cycle engines shall not exceed 900 rpm. Piston speed shall not exceed 1,725 fpm.

Recommended rpm and piston speeds are somewhat lower for two-cycle engines.

Cooling System

Generally, stationary diesel engines are jacketed and fresh water cooled by an engine-driven pusher fan and radiator. Coolers sized for 100°F ambient temperatures are common. The coolers may be vertical or horizontal, the latter used when space is at a premium. The cooling system shall be thermostat controlled and have explosion-proof jacket water heaters. Some large industrial engines warrant the use of cooling liquid to water shell-and-tube or plate-and-frame heat exchangers.

Starting System

The most common ways to start diesel engines are:

- By using 24-volt direct current (D-C) and an explosion-proof electric starter, together with several sealed, maintenance-free batteries housed in a fiberglass battery box. The system shall also include an alternator, trickle charger, voltage regulator, ammeter mounted on an instrument panel, and all appropriate switches and wiring.
- By using air start motors, which include an oiler, a strainer, a solenoid-operated air starting valve, and an air-pressure regulating valve. Air supply may come from plant air, natural gas, or nitrogen bottles. The system shall include all necessary piping and valves. Usually a 250 psig air supply will suffice.

Both air and electric starters shall be capable of three progressive starts without pressurizing the air supply or charging the batteries.

Coupling

Preference is for a flexible coupling installed between the engine flywheel and the pump shaft.

Exhaust

An exhaust pipe with a spark-arresting silencer is usually mounted on top of the engine or on top of an eventual enclosure. A stainless steel exhaust flexible joint connects the exhaust system to the engine. For personnel protection, aluminum-jacketed calcium silicate, fiberglass, or mineral wool insulation covers the exhaust system.

Governor

The use of an electromechanical governor for load sharing and speed control is common.

Barring Device

The engine manufacturer shall supply a barring device for use during pump maintenance and/or repairs.

Instrumentation

Typically, a diesel engine shall have at least the following instruments and alarms:

Instruments

- Lube oil pressure
- Air-pressure intake manifold
- Jacket water pressure
- Lube oil temperature
- Aftercooling water temperature
- Tachometer
- Jacket water temperature in
- Jacket water temperature out
- Running hours counter

Alarms and Shutdowns

The engine shall have double-pole, double-throw switches on a local and/or remote control panel for alarms and shutdowns as follows:

	Alarm	Shutdown
Low lube-oil pressure	X	X
High jacket water temperature	X	X
Low fuel-level day tank	X	
Overspeed	X	X
Excessive vibrations	X	
Low water-level expansion tank	X	
High lube-oil temperature	X	
Low starting air pressure	X (if applicable)	

Steam Turbines

Power plants and many industries with a steam supply commonly use steam turbine as pump drivers.

Industry Specifications

- *API Standard 611*   General-Purpose Steam Turbines for Refinery Services
- *API Standard 612*   Special-Purpose Steam Turbines for Refinery Services
- *ASME*   Boiler and Pressure Vessel Code, Section IX, Welding and Brazing Qualifications
- *NEMA SM 23*   Steam Turbines for Mechanical Drive Service.

Features

Consider the following features when buying a steam turbine.

Application

General-purpose steam turbines for driving pumps shall have inlet steam pressure not exceeding 600 psig at 750°F and 200 psig exhaust pressure. Turbine speed shall not exceed 6,000 rpm.

Power Requirement

The steam turbine total power output shall not drop lower than 110 percent of the rated pump brake horsepower, plus any mechanical and transmission losses at minimum inlet and maximum steam state.

Screen

The steam supply line shall have a screen to remove any solids in the stream. The correct place for this screen is ahead of both the governor valve and the combined trip and throttle valve. The screen metal shall be corrosion resistant, and the screen shall be removable without breaking any flange connections in the supply piping.

Speed Governor

This controls the turbine speed by varying the steam supply to the turbine. A series of governor-controlled valves assist the speed governor. These valves were designed for minimum throttling losses. A relay-controlled hydraulic mechanism controls the sequencing of these valves.

Provisions do exist for adjusting the settings on the governor to change the speed and/or the power output while the turbine runs. This adjustment may be done locally or by remote control from a central control room.

Overspeed Governor

The overspeed governor exists to trip the throttle valve at a predetermined speed and thus stop the steam from flowing into the turbine. The overspeed governor functions completely independently from the regular speed governor.

Trip and Throttle Valve

Usually the trip valve and the throttle valve are combined. Together with the emergency overspeed governor, they automatically stop the steam from flowing into the turbine when the turbine reaches a certain overspeed. During start-up, a manually operated valve controls the steam flow into the turbine.

Turbine Casing

The recommended metallurgy for casing parts that will contain high temperature (500°F) is carbon steel. If working pressures allow, the rest of the casing can be close-grained cast iron. Preferably, the casing and the diaphragms/blade rings are axially split for easy removal of parts during maintenance and repair. The turbine casing shall have centerline supports.

Turbine Rotor

The turbine rotors are heat-treated forged steel, dynamically balanced to run without vibrations.

Bearings

The radial bearings usually consist of babbitt-lined, axially split sleeve bearings. The thrust bearing shall, as a minimum, handle loads twice that of the design capacity. The bearings shall be either oil ring lubricated or pressure lubricated. Bearing housing end openings shall have seals to prevent lube oil contamination.

Shaft Seals

The seals on turbines with 50 psig or smaller exhaust pressure where the turbine shaft passes through the turbine casing and between stages shall be carbon rings. The shaft seals on turbines with greater than 50 psig exhaust pressure shall be a combination of carbon rings and labyrinth seals.

Lubrication

The turbine-shaft-driven lube oil pump supplies the oil for both the hydraulic turbine operation and lubrication. The pump takes suction from an oil reservoir supplied by the turbine. Some turbine manufacturers provide an auxiliary lube oil pump for start-up or emergency use.

A pressure regulator automatically controls the lube oil pump. Other auxiliary items are either single or twin oil coolers and strainers, as well as an OSHA-approved coupling guard. Turbine sections containing high-temperature steam have metal-jacketed insulation for personnel protection.

Gas Turbines

You won't commonly see gas-turbine-driven pumps, but they do exist. Large pipeline shipping and booster pumps have gas turbine drivers as well as some water injection oil field pumps. The gas turbine packages usually come complete with the following equipment:

- Gas engine
- Power turbine
- Gear box (if necessary)
- Seal and lube oil console
- Starting system

- Instruments
- Fuel system
- Instrument panel with alarms and shutdowns
- Fire protection equipment
- Weather-proof and sound-attenuated enclosure
- Package base or oil-field type skid

Fuel System

Many gas turbines can operate on a variety of fuels. Many stationary units run on natural gas or diesel oil. They may also have dual fuel systems. A typical dual fuel system may be for the use of natural gas and a distillate.

Exhaust Heat Recovery

To increase efficiency, many turbines use a combined cycle. That is, they use the heat from the exhaust gases to preheat the combustion gas. The heat from a heat recovery unit placed on top of the exhaust stack may be used for other applications, such as:

- Making steam
- Space heating
- Air conditioning
- Water desalinization

Governor

Gas turbines have either electric load-sensing governors for multiple units running in parallel or hydraulic governors for single units.

Starters

The most common starters are air or gas expansion motors. Starting gas pressures may reach 250 psig. Another option may be 24-volt D-C electric starters with battery banks and chargers.

Ratings

Gas turbine horsepower ratings are given for continuous duty at sea level with an air inlet temperature of 59°F. At different altitudes and temperatures, the ratings vary. At cooler inlet air temperatures, the efficiency increases; at higher altitudes and higher inlet air temperatures, the efficiency decreases. The gas turbine manufacturers have developed charts for de-rating the turbine efficiency. Because all gas turbine manufacturers have different designs, their rating curves also differ.

Hydraulic Drives

Electric motors in downhole applications present a potential problem. Both the motor and the electric power cable exist in a hostile environment, submerged in often-corrosive liquids. Another limitation is speed. The diameter of the pump impeller, which generates the pump head, is limited to the bore hole diameter, which in deep wells seldom exceeds 7 in. OD. A two-pole motor develops a maximum of 3,600 rpm. To produce the desired head to lift the liquid to the surface, these electric submersible pumps (ESPs) are always multistaged. More than 30 stages on an ESP is not uncommon.

Several pump manufacturers use hydraulic turbines instead of electric motors as drivers to improve downhole pump reliability and to reduce initial costs. The principle is simple: A clean, pressurized liquid stream is led through a supply pipe to the turbine connected close coupled to the pump. The liquid used to power the turbine may be generated at the surface specifically for that application or may be a side stream from another unrelated high-pressure source.

The electric turbine may spin at up to 10,000 rpm, thereby drastically reducing the number of stages needed to produce a given head. The turbine also eliminates the frequent pulling of ESPs to replace short-circuited electric motors.

Industry has been slow in accepting hydraulic drives for downhole applications, possibly because of several partially successful attempts to develop a reliable hydraulic drive. In the last few years, one of the larger British pump manufacturers designed a reliable pump coupled to a hydraulic drive.

Solar Power

The development of photovoltaic (PV) panels has made solar-powered pumps a reality. The pump drivers may be either D-C electric motors (12 or 24 v) or, by using an inverter to convert the D-C power generated by the solar panels to A-C, 125-v A-C motors. Battery banks store the power generated by the PV panels.

Common applications for solar powered pumps include:

- Homes in rural areas not serviced by an electric grid.
- Water well pumps in isolated villages in undeveloped countries.
- Irrigation projects.
- Supply for livestock troughs.
- Water circulation in swimming pools.
- Automatic bilge pumps in boats.

Chapter 14

Gears

The following standards apply to gears installed in pump packages:

- *AGMA (American Gear Manufacturers Association) 420.04, Practice for Enclosed Speed Reducers or Increasers Using Spur, Helical, Herringbone, and Spiral-Bevel Gears*
- *AGMA 421.06, Practice for High-Speed Helical and Herringbone Gear Units*
- *API Standard 610, Special-Purpose Gear Units for Refinery Services*

Pump applications often include the use of a gear train, such as:

- Parallel shaft gears
- Right-angle gears
- Epicyclic gears

Parallel Shaft Gears

The gears come in various configurations:

- Spur
- Single
- Helical
- Double helical
- Herringbone

Spur Gears

In spur gears (Figure 14.1), the teeth run parallel to the gear shaft. Spur gears are less expensive to make than helical gears, which have angled teeth. Because of the gradual contact of the teeth between mating gears, helical gears are more reliable than spur gears. A helical gear has greater face area than a spur gear, and the helix in the pinion gear gives it the capacity to produce twice the horsepower of a spur gear. Installation of spur gears in pump packages is not common.

Gears   115

Figure 14.1 Spur Gear *(Courtesy of Lufkin Industries Inc.)*

Helical Gears

Single-Helical Gear
When buying or specifying a high-speed gear, features to consider include the following:

* The gear design and construction shall be such that the gear can run continuously at full load at the operating conditions shown on the pump or gear data sheets.
* Consider the suggested service factor for parallel shaft single-stage helical (Figure 14.2), listed below for various pump configurations:

Pump	Driver	Service Factor
Reciprocating	Internal combustion	2.4
Reciprocating	Electric motor	2.0
Reciprocating	Gas turbine	2.0
Centrifugal	Internal combustion	2.0
Centrifugal	Electric motor	1.6
Centrifugal	Gas turbine	2.0

- The critical lateral and torsional speed shall be a minimum of 20 percent above or below the rated input speed.
- The low-speed gear shall be shrink fitted and preferably keyed to its shaft. The high-speed pinion gear shall be forged and machined together with the shaft. Matched gear sets shall be match-marked.
- The shaft may be either cylindrical or tapered. Usually shafts of 1½ in. and larger taper to accommodate hydraulic fit coupling hubs.
- The gear housing may be fabricated steel for smaller gears (150 hp and smaller). Gears 150 hp or larger shall preferably be cast carbon steel or cast stainless steel. The casing shall be split horizontally. The two mating surfaces shall be machined to a high finish and joined by cap screws and hex nuts.
- Cast casings shall have machined mounting surfaces, and fabricated casings shall have fabricated soleplates. The casing shall preferably have two lifting lugs and an inspection plate.
- Piping connections shall be flanged or socket welded, except for instruments and instruments valves that may be screwed. Seal-welded screwed piping is not recommended.
- Split-sleeve, babbitt-lined, steel-backed journal bearings with stabilizing features shall support both shafts at each end. Double-helical gears shall preferably have a double-acting, pivoted segmental thrust bearing on the low-speed shaft.
- Generally, the pump or the driver manufacturer supplies a common lubrication system for the pump, driver, and gear train. The lubrication shall be force feed. The lube oil piping shall be designed to spray oil when the gears come into mesh and come out of mesh.
- The bearings shall operate at a maximum of 200°F in the bearing-loaded area. Thus, with an inlet temperature of 120°F, the oil temperature rise through the bearings shall always be less than 80°F.
- The gear casing shall have a dry type filter-breather.
- All external piping shall be either ASTM A106 seamless carbon steel or minimum schedule 40 300 series stainless steel.
- Gear case vibrations should be limited to 0.5 mils, peak to peak in any direction.

Double-Helical Gears

This type of gear (Figure 14.3) has two opposed sets of helical gears with a groove separating them. The main advantage of a double-helical gear over a single one is the absence of internal thrust to the shaft and bearings. However, external thrust loads may shift the load of one helix to the opposite one, which may overload that side.

Herringbone Gears

A herringbone gear (Figure 14.4) is similar to a double-helical gear, but no groove separates the two rows of teeth. The use of herringbone gears is common where a demand for transmitting heavy loads at relatively low speeds exists.

Gears 117

Figure 14.2 Single-Helical Gear (*Courtesy of Lufkin Industries Inc.*)

Figure 14.3 Double-Helical Gear (*Courtesy of Lufkin Industries Inc.*)

118 *Practical Introduction to Pumping Technology*

Right-Angle Gears

Vertical turbine pumps driven by internal combustion engines have either a straight bevel or a spiral bevel to transmit power from a horizontal shaft to a vertical one. Straight-bevel gears have straight teeth on both the pinion and gears. On a spiral-bevel gear (Figure 14.5), the curved teeth lie at an angle. Manufacturing costs

Figure 14.4 Herringbone Gear (*Courtesy of Lufkin Industries Inc.*)

Gears **119**

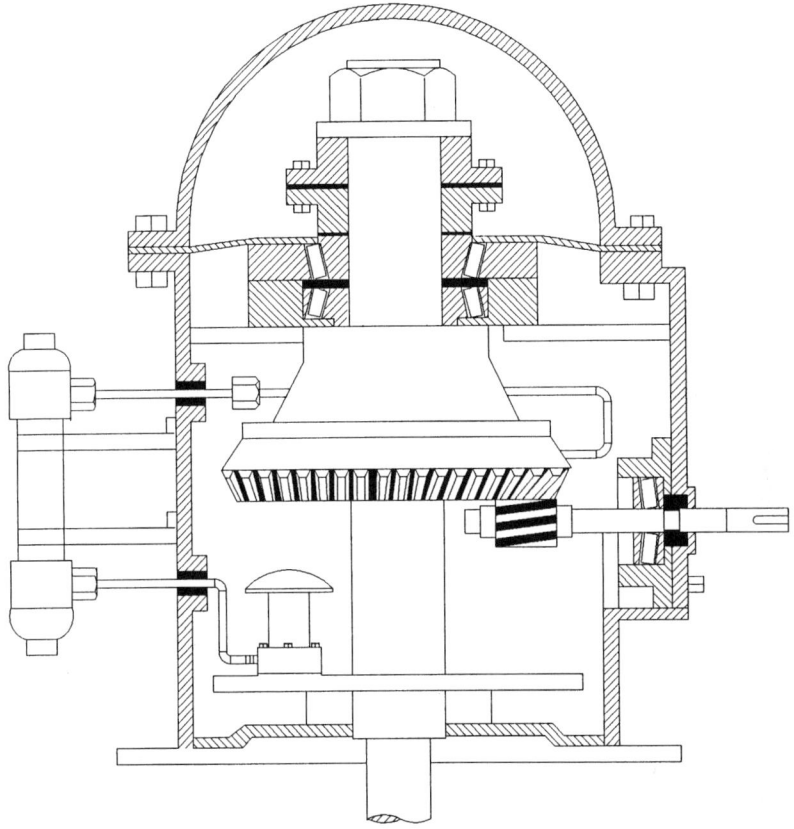

Figure 14.5 Spiral-Bevel, Right-Angle Gear

for a spiral-bevel gear are higher, but spiral-bevel gears perform better than straight-bevel gears because, at any time, two or more teeth share the loading on spiral-bevel gears.

Often, right-angle gears transmit power at the same speed as the driver and pump, because many diesel and natural gas engines run at 1,800 rpm, which is the same as most vertical turbine pumps. Do not operate this type of pump at higher speeds. However, some users prefer to run the pumps at lower speeds, and in that case the right-angle gear also functions as a speed reducer. Lubrication for this type of gear is similar to the lubrication systems used in parallel shaft gear drives.

Right-angle gears, the same as vertical electrical motors, may have solid or hollow shafts. In either case, the gears used on most U.S.-designed vertical turbine pumps

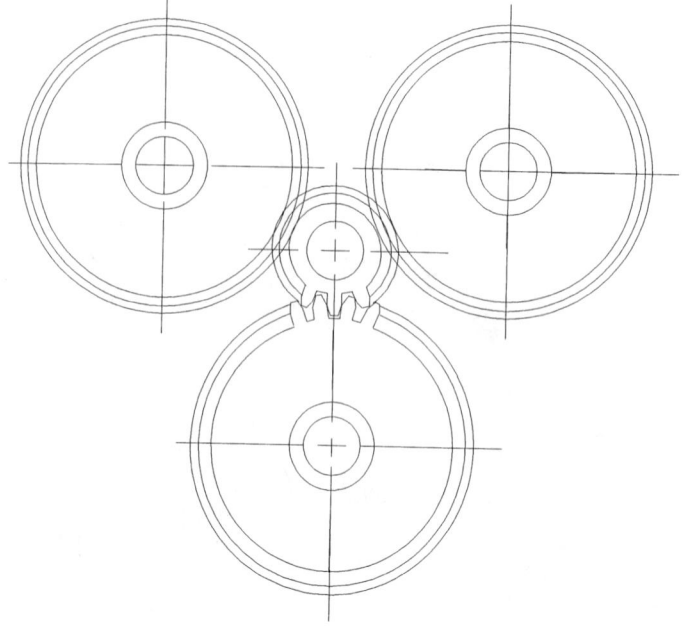

Figure 14.6 Planetary Epicyclic Gear

will have a thrust bearing at the top of the shaft. European pump manufacturers seem to prefer to incorporate the thrust bearings in their pump designs.

Low horsepower pumps also may have worm-gear drives when an installation requires horizontal change in direction between the driver and the pump.

Epicyclic Gears

Planetary-type epicyclic gears (Figure 14.6) are light co-axial gears that sometimes replace parallel shaft gears in pump packages. A sun gear, integral with the input shaft, drives planetary gears that drive the output shaft.

Chapter 15
Couplings

Types of Couplings

Couplings connect pumps and/or gears to their drivers. Types of couplings include:

- Rigid couplings
- Flexible couplings
- Lubricated couplings (encompassed in first two categories)
- Nonlubricated couplings (encompassed in first two categories)

Rigid Couplings

One type of rigid coupling consists of two flanged hubs held together by bolts and nuts. The friction between the two flanges prevents putting heavy shearing loads on the bolts. Rigid couplings do not allow misalignment between the driven equipment and the driver. The use of rigid couplings is common when connecting solid-shaft vertical motors to vertical pumps. Often users ask for adjustable rigid couplings (Figure 15.1). These are the same as regular flanged rigid couplings, with the addition of a threaded ring between the two flanges. The ring adjusts the position of the impellers in the pump bowls or casings. Most vertical pump manufacturers make their own rigid couplings. The couplings require accurate machining to avoid a potentially catastrophic misalignment.

Split rigid couplings (Figure 15.2) are axially split coupling hubs also held together by bolts and nuts. The addition of a key in the coupling and the two shafts makes the connection stronger. Split rigid couplings often connect line-shaft sections.

Flexible Couplings

Another way to transmit torque from the driver to the pump involves using flexible couplings. These couplings also allow for some misalignment between the two shafts (Figure 15.3).

Some flexible couplings use elastomers to achieve flexibility; others use various mechanical devices. Commonly used flexible couplings include:

- Pin and bushing
- Jaw

122   *Practical Introduction to Pumping Technology*

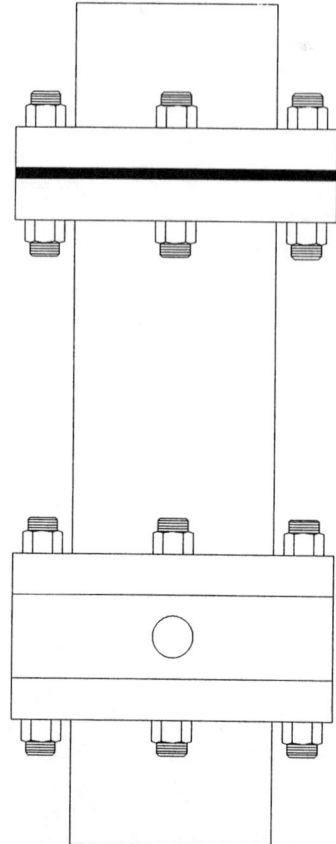

Figure 15.1 Adjustable Rigid Coupling

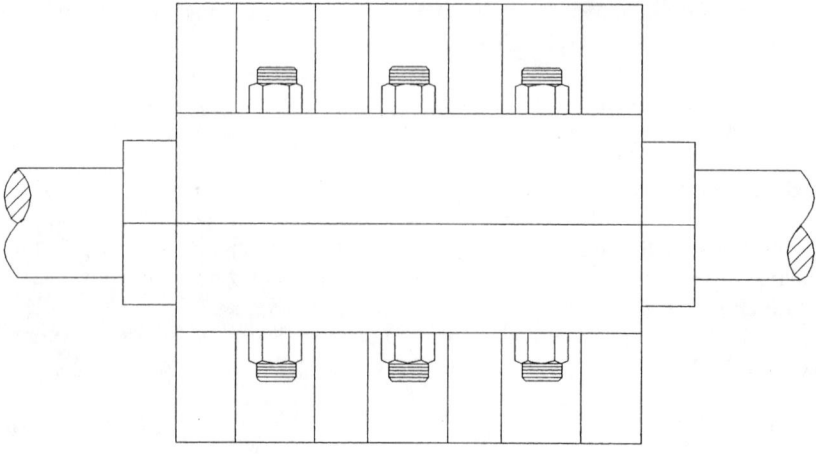

Figure 15.2 Split Rigid Coupling

Couplings **123**

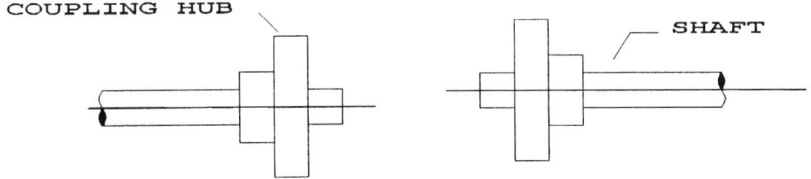

PARALLEL SHAFT MISALIGNMENT

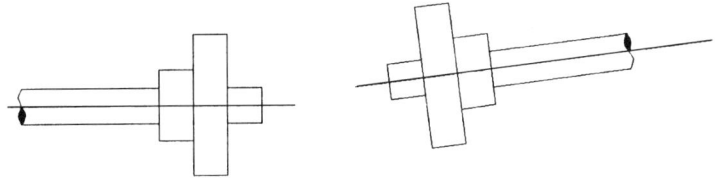

PARALLEL AND ANGULAR MISALIGNMENT

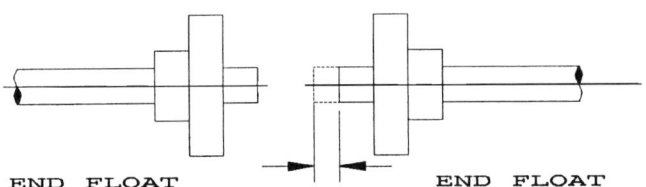

END FLOAT   END FLOAT

Figure 15.3 Shaft Misalignments

- Sleeve
- Gear
- Roller chain
- Diaphragm
- Disc
- Universal joint

Pin-and-Bushing Couplings

One of the most widely used types of flexible coupling is the pin-and-bushing coupling (Figure 15.4). This coupling uses elastomers with great pliancy to enable it to bounce back to its original form. Elastic bushings give the coupling its flexibility.

Jaw and Sleeve Couplings

These two categories of flexible couplings are clamped elastomer couplings.

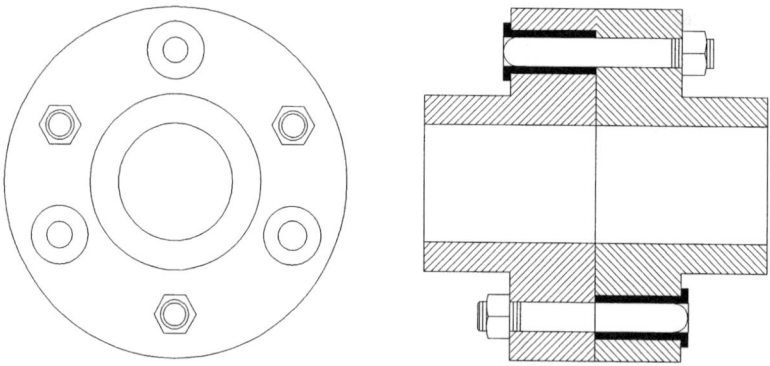

Figure 15.4 Pin-and-Bushing Coupling

Gear Couplings

The pump industry commonly uses the gear-type flexible coupling (Figure 15.5). The design includes clearance fit splines on the coupling hubs. A sleeve with mating internal gear teeth joins the hubs. There is some backlash in the spline joint. The backlash allows some misalignment of the shafts, and the backlash action causes wear on the splines. Therefore, this type of coupling needs either grease or oil lubrication.

Roller Chain Couplings

The roller chain coupling consists of two keyed hubs with sprockets cut on the mating shoulders of the hubs. A bicycle-type roller chain joins the two hubs. The clearance between the chain and the sprockets gives this coupling flexibility. Inexpensive, low-speed pump packages commonly have this type of coupling.

Industry is turning more and more toward nonlubricated couplings. The most popular of these are the diaphragm and disc-type flexible couplings.

Diaphragm Couplings

Diaphragm-type couplings consist of a curved, flexible steel disc on each hub. Either a spacer or a center member separates the two halves. Usually, the hubs and the center member consist of carbon steel, and the flexing elements are AISI type 300 series stainless steel. The thickness of the diaphragms defines the torque applied to the coupling.

Couplings   **125**

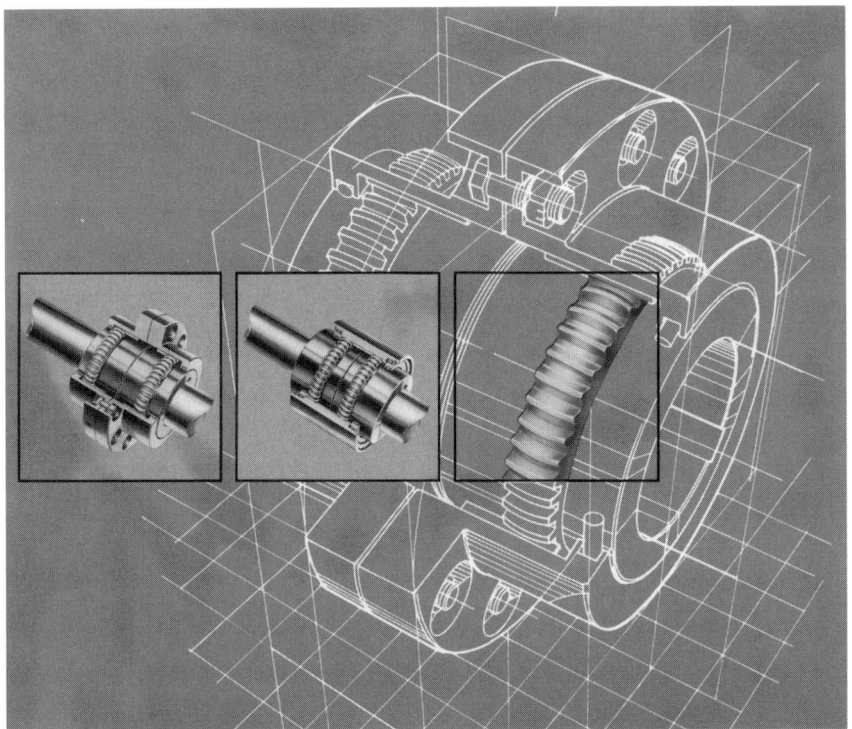

Figure 15.5 Gear-Type Coupling (*Courtesy of Ameridrives International*)

Disc Couplings

The disc coupling (Figure 15.6) works on the same principle as the diaphragm type, but with stacked circular discs to provide flexibility. More discs provide higher torque.

A coupling used with a pump that has a mechanical seal usually has a spacer between the two halves. By loosening and dropping the spacer, maintenance personnel can easily replace the mechanical seal without disturbing any piping.

Universal Joints

The recommended coupling for connecting a right-angle gear to an internal combustion engine is a universal joint attached to a tubular shaft (Figure 15.7). Offset the driver shaft and the gear shaft between 3° and 6°.

126 *Practical Introduction to Pumping Technology*

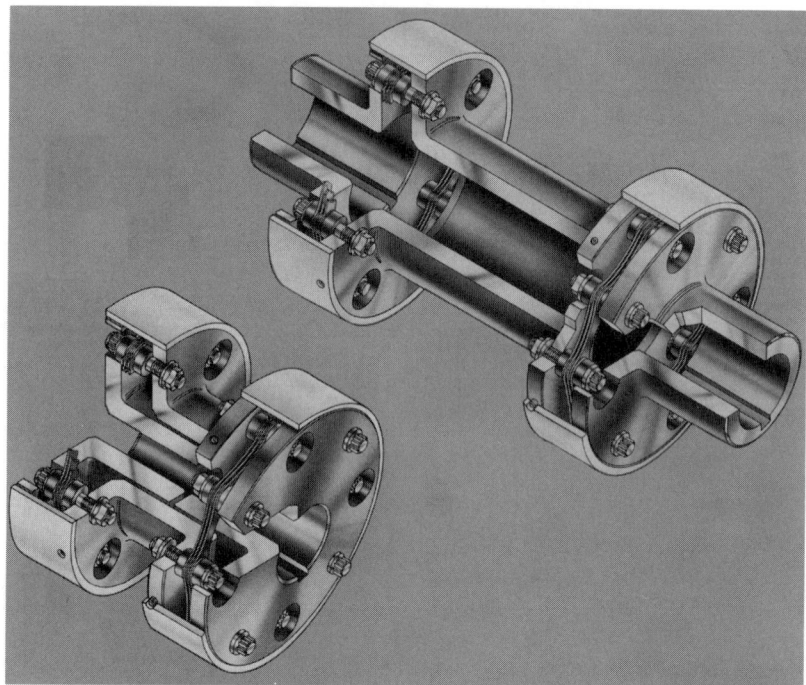

Figure 15.6 Disc-Type Flexible Coupling (*Courtesy of Ameridrives International*)

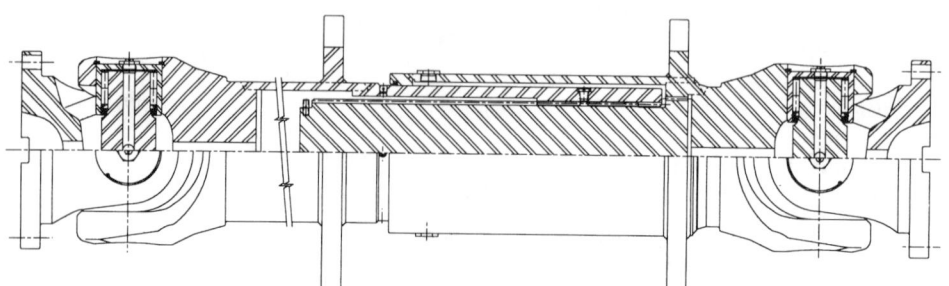

LONG TRAVEL CAPABILITY WITH EXPANSION AND CONTRACTION FLANGES

Figure 15.7 Universal Joint (*Courtesy of Ameridrives International*)

Typical Service Factors

Pump Type	Service Factor
Motor- and Turbine-Driven Pumps	
Centrifugal	
General liquid service	1.0
Boiler feed pumps	1.0
Sewage pumps	1.5
Slurry pumps	2.0
Rotary	1.5
Reciprocating	
Double-acting pumps, up to two cylinders	2.0
Single-acting pumps, three or more cylinders	2.0
Internal Combustion Driven Pumps	
Engines with eight or more cylinders	2.5
Engines with six cylinders	3.0
Engines with four cylinders	3.5

For engines with less than four cylinders, consult the coupling manufacturer.

Chapter 16

Pump Controls

Pump operations usually control only one variable: flow, pressure, or level. All pump control systems have a measuring device that compares a measured value with a desired one. This information relays to a control element that makes the changes.

Ways to vary flow in the pump discharge piping include throttling, using a bypass, or using speed controls. A throttle valve and/or a minimum bypass line usually controls centrifugal pump flows. The use of variable speed drives is more limited but is increasing.

By changing cylinder volume and/or pump speed, you can control reciprocating pumps.

Control Valve Types

A pump system uses two basic types of control valves:

* On-off valves
* Modulating valves

On-Off Valves

You can control a pump in the simplest way with stop/start relays and manually operated suction and discharge block valves. The block valves are either closed or opened. The discharge block valve may also be used as a manually adjusted throttling valve.

Valves for on-off services are usually gate, ball, or butterfly valves. Sometimes a plug valve is appropriate if the pump user plans to use the discharge block valve regularly as a control valve.

The pump user may obtain control with manually operated valves or sophisticated microprocessors. Economics dictate the accuracy and complication of a control system.

Modulating Valves

These valves control pump discharge and/or bypass flow rates. They may also control ancillary flows, such as the flush rate for mechanical shaft seals. Control valves are ball, globe, butterfly, or plug valves. All control valves have a valve body, a variable orifice, and an actuator that controls the orifice.

An actuator controls the modulating valve. A measuring device, such as an orifice plate, measures the flow. This information then relays to an actuator that either closes or opens the valve as required. The measured variable is usually minuscule. Transducers may convert the signal into a measurable electric signal. This signal may then convert into air pressure.

Pumping systems often use microprocessors to monitor the pump process, driver, and auxiliary equipment. Examples include PLCs (Programmable Logic Controller), SCADA (Supervisory Control and Data Acquisition), and DCS (Distributive Control System).

Capacity Control

Control Systems

A pump exists to deliver a predetermined liquid flow at a certain head. A flow control valve, a variable speed drive, or a combination of both may achieve this.

The open loop, or feed forward system, is one automatic control system. It controls both flow and pressure. Setting a value for a control valve allows a predetermined amount of liquid to flow. Constantly measured deviations from this value correct and adjust the valve's modulating element to the changed conditions. Operators control the input variables. For instance, an input variable may be a change in instrument air pressure that will affect actuator travel in an air-operated control valve. Open-loop systems, although simple and quick to respond, are error-prone to downstream condition changes.

The closed loop, or feedback, provides a more accurate automatic control system. A predetermined value compares and measures the output variable. If the comparison shows a difference between the two values, the pump speed will automatically change, or the valve setting will adjust to correct the error.

The primary devices that measure fluid flow by differential pressure are:

- Orifice plate
- Venturi tubes
- Flow nozzles

Discharge Control Valves

A flow sensor picks up a change in flow conditions, such as a demand for more capacity. A signal transmits this information to the motor control center via a flow transducer. The motor control center then sends a signal to the actuator, telling it to open the valve orifice. (See Figure 16.1.)

Large discharge control valves, 10 in. and larger, are usually ball valves for high-pressure applications and butterfly valves for low-pressure ones. Globe and plug valves better suit smaller pipe sizes. Valve manufacturers also market low-noise control valves. They achieve noise reduction by making the liquid flow through numerous small passages. The pump user should use these valves with caution because they have a tendency to plug.

Large control valves, 6 in. and larger, require block valves, both upstream and downstream, with a bypass line to keep the pump installation on-stream during maintenance and testing. Spur gear operated bypass valves do not hold up well to

SELF-CONTAINED BACK PRESSURE REGULATOR

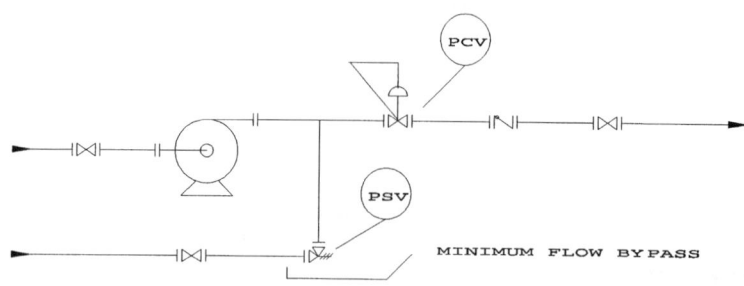

PRESSURE REGULATOR WITH EXTERNAL PRESSURE TAP

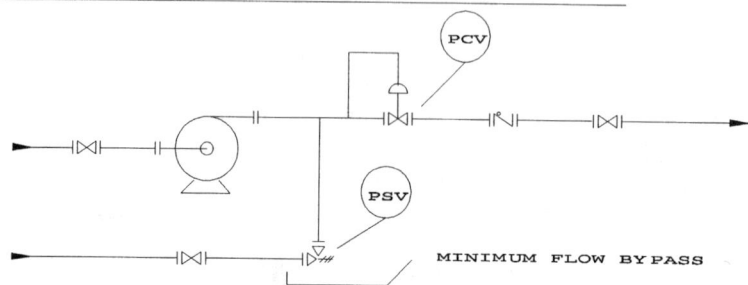

Figure 16.1 Pump Pressure Control

strong closing torque and should be avoided. There is no need for bypass valves with pumps in intermittent services.

Electric motors, rotary air motors, or hydraulic pistons drive actuators. When a system requires a longer stroke or high-speed response, the use of a pneumatic-piston-activated control valve is recommended. Electrohydraulic valve actuators are common when instrument air or gas is unavailable. Very large ball valves often have this type of actuator.

Variable Speed Control

Changing the pump speed influences a centrifugal pump's capacity and head. On the other hand, speed variations affect only the capacity of positive displacement pumps. Speed control is attractive with large-flow, high-pressure pumps, where it can offer substantial energy savings compared with throttle control. Speed control is also used with pumps that require large flow variations, such as shipping pumps, or with water flood applications, where the injection wells' injectivity index increases gradually. Often, the alternative is to initially destage these pumps, if multistaged.

Gas Turbines. These are used as pump drivers, mainly with pipeline pumps, water injection pumps, or shipping pumps. The turndown capabilities of a split-shaft gas turbine can reach 50 percent of full speed, though 80 percent is more common. Either a manually set or automatic governor may control the speed, maintaining constant flow and discharge pressure under a wide range of operating conditions. The absence of dangerous vibrations and small, unbalanced inertia forces provides an advantage of gas turbines in offshore applications.

Internal Combustion Engines. The use of engines fueled by gasoline, diesel, or natural gas is frequent where electricity is not readily available. Speed may be controlled manually or by an automatic governor.

Hydraulic Drives. A flow of liquid powers a hydraulic drive by turning an impeller that drives a runner. The runner shaft drives the pump. Liquid volume variations between the impeller and runner adjust the speed. The adjustments may be done manually or by automatic control devices. The fluid that drives needs cooling. Oil may be used often as the driving media, but a British pump manufacturer uses part of the discharge fluid from bore hole pumps to power the actual pump.

Variable-Speed Electric Motors. This is actually an electromagnetic drive that operates on the eddy current principle, as follows:

The motor has two independent rotating assemblies enclosed in a common casing. The drum, the outer assembly, connects to the input shaft. The field rotor connects to the output shaft. Magnetic flux between the two rotors creates the torque. The field excitation controls the drive electronically.

Pump users employ variable-speed electric motors when the operation calls for large capacity and head variations over a long period. Variable-speed electric motors have a speed range from 0 percent to 100 percent, which may reduce power costs.

Variable-Speed Hydraulic Couplings. These couplings integrated with parallel shaft gears are sometimes used in response to a demand for wide flow and head variations. The driver is usually a constant-speed electric motor. Some power loss in the drive offsets eventual power savings at reduced loads.

Other Variable-Speed Drives

- Eddy current variable-speed couplings
- Belt drive is common in smaller reciprocating pumps. To vary pump speed, the pump user needs to change the belt sheaves—but not, of course, while the pump is running.

Minimum Flow Bypass

In positive displacement pumps, the bypass functions as a pressure relief system and as capacity control. A minimum flow bypass in a centrifugal pump system mainly exists to protect the pump from overheating. It may also be used to regulate process flow, but it is not economical to do this continuously. To protect gradual heat buildup in the pump, the minimum flow bypass line shall be led as far upstream from the pump suction as practical.

One of the prerequisites of a bypass valve is to dissipate the discharge pressure before the bypass line enters the upstream process stream. Cage-type plug valves and drag valves accomplish this. Minimum flow valves must be tight shutoff valves. The correct place for a minimum flow bypass is downstream of the pump's discharge nozzle and upstream of the discharge check valve. When several pumps operate in parallel, each pump shall have its own bypass line.

Liquid Level Control

Regulating pump flow by the liquid level in a storage tank or a process vessel is common (Figure 16.2). This is done by:

- Mechanically operated valves activated by floats
- Float-activated, pilot-operated valves
- Diaphragm motor valves activated by a displacement-type controller or floatless level switches.
- Pressure switches

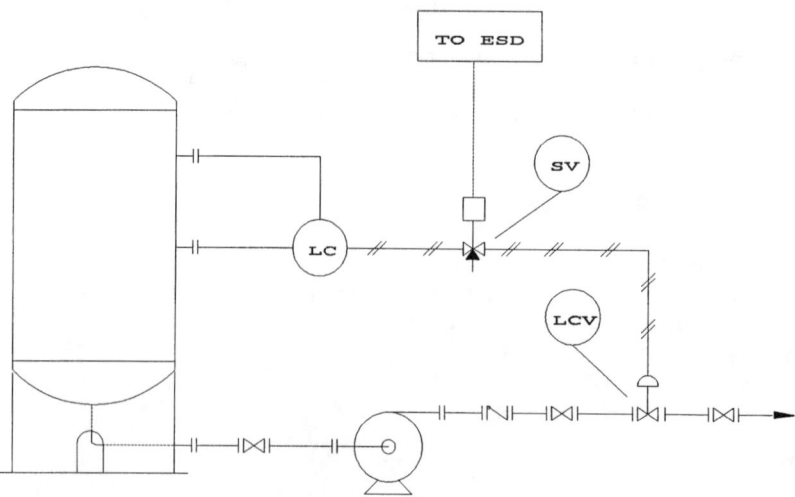

Figure 16.2 Pump Level Control

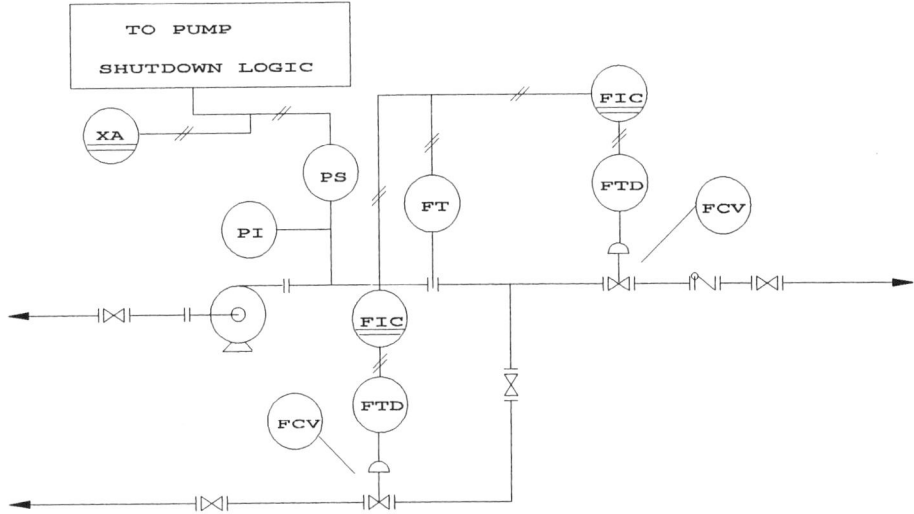

Figure 16.3 Pump Flow Control

On-Off Control

Small sump, sewer, and process pumps often have on-off control. The electric motor manufacturer limits the number of starts per day to prevent the motor windings from overheating. On-off control may be manual with push buttons, or automatic with transducers and/or transmitters that convey signals from a level controller to the motor control center.

Modulating Control

Another way to use level control is to send signals back from a level controller to a modulating control valve in the pump discharge line.

Pressure Control

Safety relief valves (PSV) are necessary when the pump shutoff head exceeds the maximum working pressure of the discharge piping or that of the downstream equipment. The capacity of a pump discharge pressure relief valve shall equal the capacity of the pump at the point on the H-C curve indicating the relief valve setting.

Relief valves discharging into a closed system shall avoid discharging directly into the pump suction line to avoid overheating the pump through recirculation of the liquid that is picking up heat from the pump. When the process fluid is very viscous, the pressure relief valve's inlet and outlet shall be heated. Sometimes companies install rupture discs upstream of a relief valve to prevent permanent contact between aggressive fluids and the relief valve.

Safety relief valves are automatic pressure-relieving devices activated by static pressure upstream. They are either rapid full-stroke pop-action-type or valves that open gradually in proportion to increased pressure.

Surge Control

Surge, or water hammer, may occur in the pump discharge piping when a valve slams shut some distance away from the pump. An emergency pump shutdown can also cause a surge in the pump suction line. Both happenings may cause severe damage to the pump and piping. To prevent surge damage, gas-loaded surge drums and/or specially constructed surge relief valves with reactions in the low millisecond range are used.

Control Selection for Positive Displacement Pumps

Capacity Control

A bypass line, changes in pump speed, and variation in piston stroke length help control the positive displacement pump's capacity. Positive displacement pumps use either pistons, plungers, gears, vanes, screws, or diaphragms to move liquids. These pumps are not kinetic, as are centrifugal pumps. They do not need velocity to create pressure. A characteristic of positive displacement pumps is that capacity is a function of speed and is relatively independent of the discharge pressure. The pump may run from stall to maximum speed without affecting the discharge pressure. This makes discharge throttling as a means of flow control impractical.

A bypass valve helps when starting a reciprocating pump. These pumps tend to self-prime. Each of the pump's cylinders function as a separate unit running in series with the other cylinders. Possibly only one cylinder in a multiplex pump can be primed while the other cylinders are vapor locked. Normally, air fills the pump cylinders during installation and maintenance. On start-up, gas or air is often drawn into the pump's suction side. Opening the bypass valve during the first pumping cycles keeps the discharge pressure low, thus helping the pump purge the air from the cylinders. That way, it will be fully primed when exposed to the full discharge pressure.

Another advantage of a bypass is that by keeping it open, on start-up, starting torque may drop as low as 75 percent less than full starting torque. Also, if a pumps starts to run rough during normal operation, air or gas may have entered one or more cylinders. Opening the bypass valve until the pump runs smoothly remedies this situation.

Safety Relief Valves and Block Valves

A positive displacement pump shall either have an integral safety relief valve or one installed in the discharge piping. The valve size shall allow passage of full pump capacity at a pressure not exceeding 110 percent of the pressure relief valve's opening pressure. To avoid feeding the gases freed through the relief valve back into the pump, the pressure relief line shall be piped back into the suc-

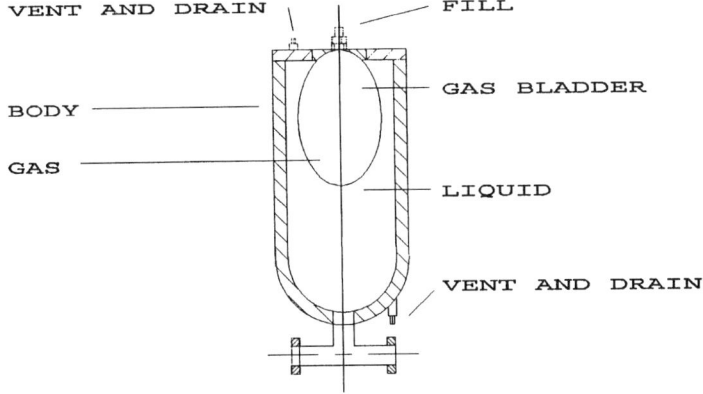

A. BLADDER-TYPE DAMPENER

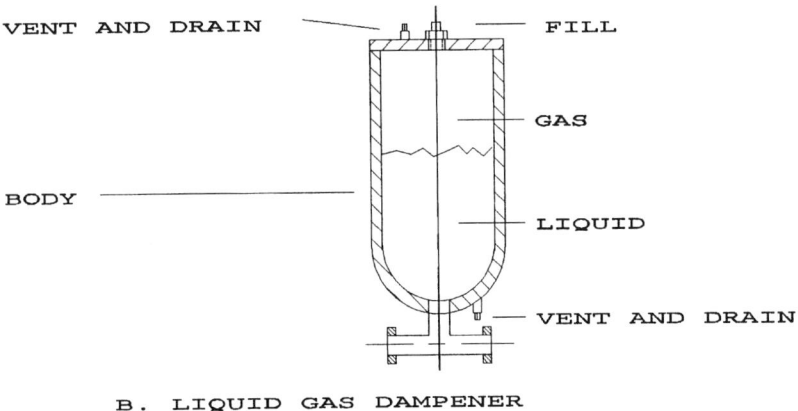

B. LIQUID GAS DAMPENER

Figure 16.4 Typical Pulsation Dampeners

tion vessel or an independent sump line. Block valves on the pump discharge side shall be fully open to avoid throttling the flow.

Variable Speed Control

Positive displacement pump speeds may vary greatly. Pulleys and V-belts control speeds in smaller pumps (50 hp and smaller). Throttling internal combustion engines or downturning steam and gas turbines are other methods of speed control, as are variable-speed electric motors and hydraulic couplings.

Stroke Adjustment

Piston stroke adjustment is one means to control the capacity of diaphragm-type chemical injection pumps. Stroke adjustments are also commonly used with reciprocating piston and plunger pumps.

Pulsation Dampeners

In reciprocating pumps, the flow in the discharge and suction piping is not constant. The flow accelerates and decelerates a number of times for each turn of the crankshaft. The pulsations may cause failure in the pumping system components and may also cause the pump to cavitate. Pulsation dampeners, also called pulsation bottles, on both the suction and discharge sides help avoid these rhythmic pulsations (Figure 16.4).

Chapter 17
Instrumentation

Instruments

Protective instrumentation on rotating equipment presents a question of pump size, application, and amount of operator supervision. Pumps under 500 hp often come with no instrumentation except pressure gauges and perhaps temperature indicators. These gauges may be on the pump itself or in the suction and discharge lines immediately upstream and downstream of the suction and discharge nozzles.

Temperature Detectors

Dial thermometers or three-wire platinum resistance temperature detectors (RTD) monitor temperature variances that may indicate a possible source of trouble. Stator winding coils on electric motors 250 hp and larger often have RTD elements. Two RTDs per phase is standard. Journal bearings in pumps and electric motors usually have one RTD element in the shoe of the loaded area (Figure 17.1). Tilted-pad thrust bearings have an RTD element in the active, as well as the inactive, side.

Dial thermometers monitor oil thrown from bearings. They are often installed without a thermowell. Sometimes temperature detectors also monitor bearings with water-cooled jackets to warn against water supply failure. Pumps with heavy wall casing may also have casing temperature monitors.

Vibration Monitors

Pumps and motors 1,000 hp and larger may have the following vibration monitoring equipment:

- A seismic pickup with double set points installed on the pump outboard bearing housing
- Proximators with X-Y vibration probes complete with interconnecting co-axial cables at each radial and thrust journal bearing
- Key-phasor with proximator and interconnecting co-axial cables

Annunciators, Alarms, and Shutdowns

Annunciators and alarms provide operators with warnings of abnormal conditions that, unless corrected, will cause pump shutdown. Systems designed to protect

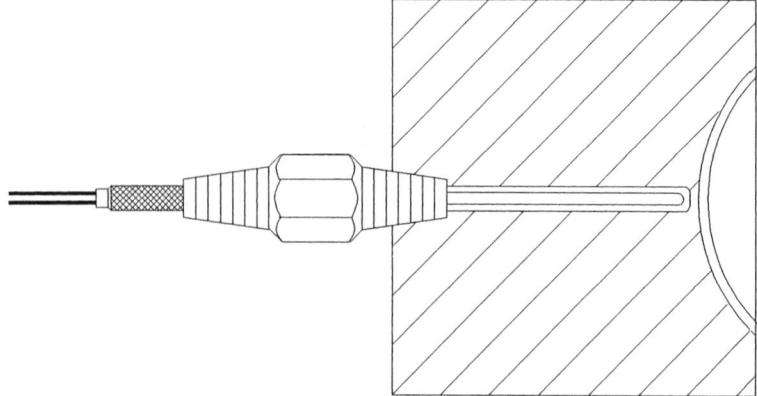

Figure 17.1 Thermocouple Installation in Journal Bearing

equipment in critical service should be as failsafe as possible. That means the use of redundant circuits. It may also mean installing an emergency generator or a battery bank for protection against full or partial power plant failure. Air stored in air receivers and air surge drums provides a safe shutdown where pneumatic control systems are used.

Annunciators used for both alarm and pre-alarm have both visible and audible signals. To prevent operators from starting a pump before correcting the conditions that caused the shutdown, pumps have shutdown permissives.

Functions

Mechanical Seal Leak Detection

Pressure switches and alarms may be used to announce tandem seal leakage. The same arrangement may also announce single seal leakages if the single seals have auxiliary throttle bushings. Leaking on double mechanical seals with a buffer fluid from an outside source is difficult to detect in anything short of catastrophical failure. However, a combination of level and pressure switches may prove effective. (See Figure 17.2.)

Pump Overheating

Pump manufacturers specify the minimum safe continuous flow for each pump model and size. Below that minimum flow, the pump will overheat and cavitate. An automatic bypass will solve eventual minimum flow problems.

Instrumentation

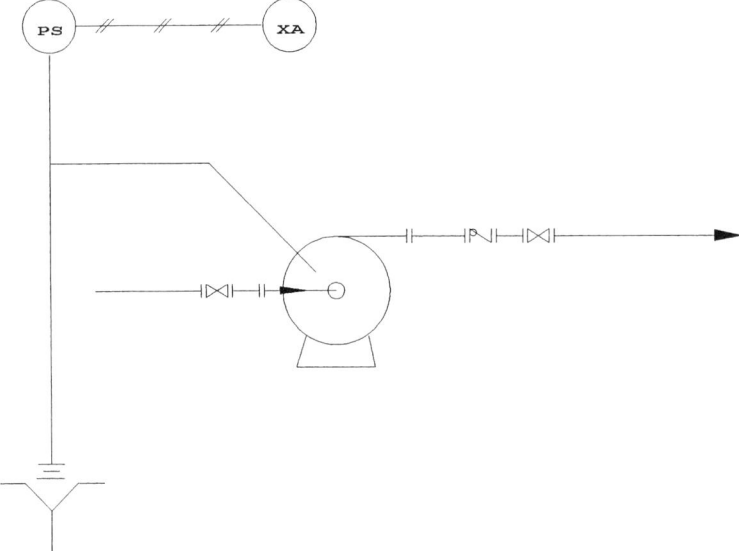

Figure 17.2 Mechanical Seal Leak Detection

Loss of Suction

If the pump suction pressure drops so the NPSHA is less than the NPSHR, the pump will start to cavitate, with possible severe damage to the pump. Often, a clogged suction strainer causes a suction pressure drop, but any number of reasons may exist. Flow or pressure switches in the suction line will protect the pump by providing an alarm or a pump shutdown when flow and pressure diminish.

Electrical Area Classification

Most panel-monitoring instruments are in unclassified areas. The enclosures for instruments in a Class I, Division 1, Group D area shall be Nema 7 (explosion proof). The enclosures for Class I, Division 2, Group D areas shall be Nema 7 or Nema 4X.

Chapter 18

Documentation

This chapter does not include commercial documentation, because every company has its own particular system. The technical documentation should include the items shown in Appendix 5 as the minimum requirement. The end user and vendor should agree on the number of copies for each document.

The buyer also has to set up a system to handle vendor documentation. At least the plant engineer, project engineer, and plant foreman should review the documents. A common practice is to make a rubber stamp with a place for each individual to initial the documents. The stamp shall also have the following words:

- Approved
- Approved as noted
- Not approved, resubmit

The vendor drawings should have the following stamps on them when received by the buyer:

- For approval (when first sent to the client)
- Approved for construction
- Approved and certified (final accepted documents)

The buyer may consider some of the following points when preparing his or her vendor data instruction sheet:

- Curves for small, inexpensive pumps or pump packages need only be standard curves, with the curve for the proper impeller trim plotted on the standard curve sheet. Dimensional drawings can also be standard drawings with a certified stamp added.
- The review of performance curves, dimensional drawings, and data sheets is the buyer's last chance to correct any mistakes in his or her specifications or in the vendor's final design. After the buyer approves the drawing, the vendor starts fabricating the package.

The next step is inspection of the finished product. The inspector will have difficulty rejecting a pump package based on a design error not caught during the drawing approval cycle.

The performance curves should be checked for the following:

- That the rated point follows specifications
- That the end of the curve performance follows specifications. Usually, the end of the curve should be at least 120 percent of the flow rate at the best efficiency curve.
- That the NPSHR is acceptable not only at rated flow, but at maximum and minimum flow rates also
- That the shutoff pressure for pumps designed to run in parallel follows specifications.

The data sheets should be checked for:

- The correct metallurgy. Often, pump vendors substitute specified material for similar material that they regularly stock. Do not, after checking the compatibility of the vendor's standard material, insist on the original material. The vendor will refuse or charge an extraordinary price for pouring a special heat for a few casings and impellers.
- The correct mechanical seals or packing. If the pump vendor recommends different seal facing than specified, consult with seal vendors.
- The correct couplings. If, for instance, the buyer's specification calls for a Metastream M Series coupling or its equal, and the vendor quotes another brand of coupling, check with the coupling manufacturers for compatibility.
- The right size driver. Recalculate the horsepower requirement for the pump's entire flow range and check that the quoted horsepower will meet the requirement, including the specified safety margin. A saving in the safety margin may lead to a smaller frame-size motor. The resulting cost difference may make this vendor the low bidder.

Engineering often prepares plant layout and pump foundations based on preliminary dimensional drawings. With ANSI pumps, this is not a problem because these pumps are standardized. However, with larger packaged pumps, revised dimensions may become a problem. Too large a pump may need special highway permits, or the package may not fit into the space allocated to it in the buyer's plant layout.

Changed anchor bolt holes may also cause problems. Usually, this only involves the buyer changing his or her foundation drawings, but the foundation may already have been poured and the anchor bolts grouted in place. Chipping out the anchor bolts and regrouting them is not serious, but it is vexing. Try to catch dimensional errors on the vendor drawings.

The buyer should also take care regarding paint. Pump manufacturers and packagers like to quote their own manufacturers' standard paint and painting procedures. These may be totally inadequate for the intended package location. For instance, packages placed outdoors on offshore production platforms require three-coat marine paint. Standard manufacturer's paint may peel off even before pump start-up. Insist on the correct paint and painting procedures. It will cost more but will avoid future embarrassment.

Chapter 19

Inspection and Testing

General Inspection

Inspection procedures may range from a full-blown witnessed performance and NPSH string test of an 8,000 hp boiler feed pump at the pump manufacturer's testing facilities to a cursory visual inspection at the buyer's shop of a 1 hp water pump. Table 19.1 shows a typical inspection checklist that any major company could issue.

Table 19.1
Inspection Requirements for Centrifugal Pumps

1	2	3	
	b	b	Radiograph inspection of impeller
	a	a	Liquid penetrant inspection of impeller
b	b	b	Liquid penetrant inspection of casing repairs
	c	c	Impact test
	b	a	Hydrostatic test
	b	a	Performance test
c			Certified performance test and NPSH curve
	b	c	Mechanical run test with vibration measurements
c			Manufacturer's standard mechanical run test
	c	c	Noise level test
	a	a	Dismantling in case of failed mechanical run test
c	c	a	Balancing of rotating assembly
	c	c	Hardness test for impeller and casing welds*

Key:
1 = Centrifugal utility pumps < 50 hp or < 250 psig discharge pressure
2 = All centrifugal pumps > 50 hp or > 250 psig discharge pressure
3 = Process pumps < 50 hp or < 250 psig discharge pressure

a = Visual inspection by company's inspector
b = Certified records examined by company representative
c = Certificates and data supplied by vendor

*If H_2S is present in fluid

The buyer must determine the inspection procedures he or she wants. If a witnessed test is necessary, the inspector should be instructed to perform the following tasks:

- Visually inspect castings
- Check welding procedures
- Visually inspect all parts before they are assembled. The castings shall not show signs of repair by peening, plugging, or brazing.
- Check running clearances on vendor's inspection records
- Check impeller and rotor balance as recorded on vendor's inspection records
- Check pump package dimensions
- Witness tests

Hydrostatic Test

Hydrotest the pump casing at 1.5 times its design pressure. The gaskets used for this test should perfectly match the gaskets used in the pump. Leaks past the gaskets are not acceptable. Discard the test gaskets after the test.

If any leakage or pressure drop occurs during the test, repeat the test after making necessary repairs. Test pumps whose medium has a specific gravity of 0.7 or less with kerosene.

Use dye penetrant solution for the hydrotest if the casing pattern is new or changed, and/or if the pump media will consistently remain above 500°F. Any dye-penetration test shall adhere to *ASTM E165*. If the buyer does not require a witnessed hydrotest, the vendor shall submit its test data for client review.

Performance Test

The buyer may insist on a performance string test, which shall include the actual driver, instrumentation, and eventual gear (see Figure 19.1 for setup). Because the vendor's lack of the electric power required may render this test inconvenient or impossible, most companies allow the vendor to use a shop driver. The shop driver often consists of a hydraulic turbine instead of an electric motor. Preferably, the test shall run at rated speed, but the vendor may run the test at other speeds. In this case, the vendor should submit the conversion-to-rated-speed calculations to the buyer for approval before starting the test. The actual string test will then occur at start-up.

The manufacturer usually will use water during the performance test. If the real pump medium is viscous, the manufacturer shall make the necessary correction, according to the latest issue of the *Hydraulic Institute Complete Pump Standards*.

The vendor often submits a list of the test equipment it will use. Upon request, the vendor shall submit calibration data for this equipment. Typical instruments include:

- An orifice with a differential pressure gauge to measure the flow rate
- Digital tachometer to measure pump rpm

- Double-element watt meter to measure power
- Pressure gauges to measure pump head
- Vibration monitor
- Temperature transmitter and RTD probe
- Sound-level meter

The performance test shall be combined with the mechanical run test, which shall last a minimum of four hours. The inspector shall check the following data, which should match the actual performance curve:

- Head against flow
- Power requirements at different flow rates
- Efficiency at different flow rates

The buyer shall select at least five flow rates for inspection. These should include the following points:

- Shut-off
- Minimum flow as specified by vendor
- A point between rated and minimum flow
- Flow at best efficiency point
- Rated flow
- 110 percent of rated flow

At each point, the inspector shall check the following:

- Pump shaft vibration
- Shaft axial displacement
- Suction pressure
- Discharge pressure
- Flow rate
- Driver input power
- Noise level

If there is a gear in the pump package, the inspector shall check the gear contact pattern after the conclusion of the performance test. Gear contact shall be consistent. A contact area of less than 75 percent will affect gear performance.

After the performance test, it is customary to open multistage and barrel pumps to inspect the rotor assembly. After draining the test liquid, the test crew shall dry all the pump components.

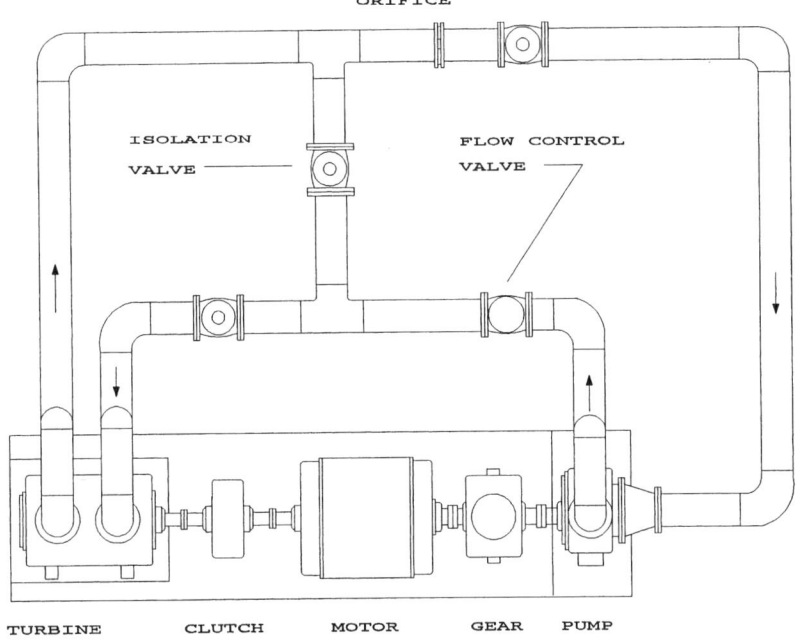

Figure 19.1 Typical Performance Test Setup

NPSH Test

It is a good idea to run a NPSH test on pumps 100 hp and larger where the NPSHA is close to the pump's required NPSH. The suppression test may determine the NPSH at 3 percent differential head loss or at 0 percent loss if the pump is in critical service with variable flow. The NPSH shall be plotted at the following flow rates, as a minimum:

- Minimum continuous flow
- Midway between minimum flow and rated flow
- Rated flow
- 110 percent of rated flow

Chapter 20

Installation and Operation

Installation

The pump purchase requisition shall include a request for vendor installation procedures. Even if the delivered pump package includes these instructions, the buyer shall consider the following items:

1. Upon receipt of the pump package, the buyer shall inspect the package for any signs of damage during transport. He or she shall also ascertain that the package conforms to the bill of lading.
2. All lifting shall be per vendor's instructions. Do not use slings around nozzles or other openings not designed for lifting.
3. Remove dirt, grease, and oil from equipment feet, or bottom of skid or baseplate. Vendor shall provide protection from damage and corrosion for these parts during transportation and storage.
4. Preferably use type 316 stainless steel, ¼-in. thick leveling shims when positioning the pump package.
5. Pump bases with four anchor bolts shall have a set of shims next to each bolt.
6. Pump bases with six or more anchor bolts shall have U-type shims around each anchor bolt.
7. Do not use wedges.
8. As a general rule, the installation tolerances shall be:
   - ⅛ in. in any direction of horizontal displacement
   - ± ⅛ in. elevation
   - ½° for face alignment in vertical and horizontal plane
9. Do not level or align the pump package by tightening bolts.
10. The largest flange, usually the suction flange, is the position selected for aligning the equipment.
11. If the pump and driver lie on the same baseplate or skid, disconnect the two before leveling the equipment.
12. Disconnect the coupling between the pump and the driver before positioning and leveling the equipment.
13. Make up the coupling after positioning and leveling the equipment, using the vendor's tolerances as a guide. Do not dowel the bases until hot aligning the package, if required.
14. Proceed with grouting of the equipment. On offshore installation, the pump skids and/or baseplates are often seal welded onto the steel deck.

15. Use a dial indicator and reverse indication for parallel and angular coupling alignment (Figure 20.1). The tolerances shall be within ±0.005 in. for both rigid and flexible couplings. These are reasonable industry standards. Use these unless Engineering specifies stricter ones.

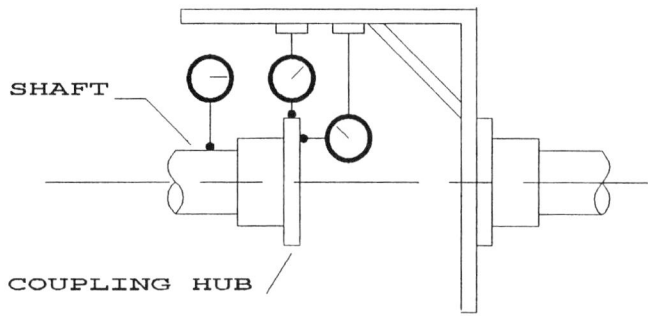

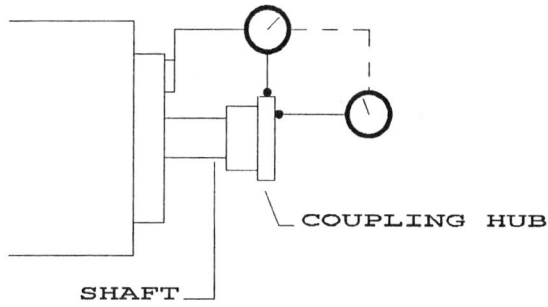

Figure 20.1 The Use of Dial Indicator to Align Pump

16. According to various sources, misalignment causes between 50 and 70 percent of all pump bearing failures, so the buyer may want to use stricter alignment tolerance and a laser indicator to get more accurate readings. To ensure proper alignment:

    - Hot check the shaft alignment after the initial full run to confirm the accuracy of the estimates for differential expansion between the two shafts.
    - Tighten the anchor bolts before connecting the piping.

Piping and Valves

Before start-up, Engineering shall check the discharge and suction piping against the following list:

1. Do not support piping on pump nozzles. Piping shall have independent support and shall not put any strain on the equipment.
2. Avoid piping configurations that might trap air bubbles, such as reverse slope or concentric reducers. Provide enough venting when in doubt. (See Figure 20.2.)
3. Place eccentric reducers with the straight face up.
4. The suction piping immediately upstream of the pump suction nozzle shall be straight, and the length shall be a minimum of 5 times the diameter of the suction nozzle.
5. Use temporary strainers during start-up and initial operation. These strainers are often cone types, with the cone pointing upstream. Remove temporary strainers afterward, because forgotten temporary strainers may cause a lot of future problems.

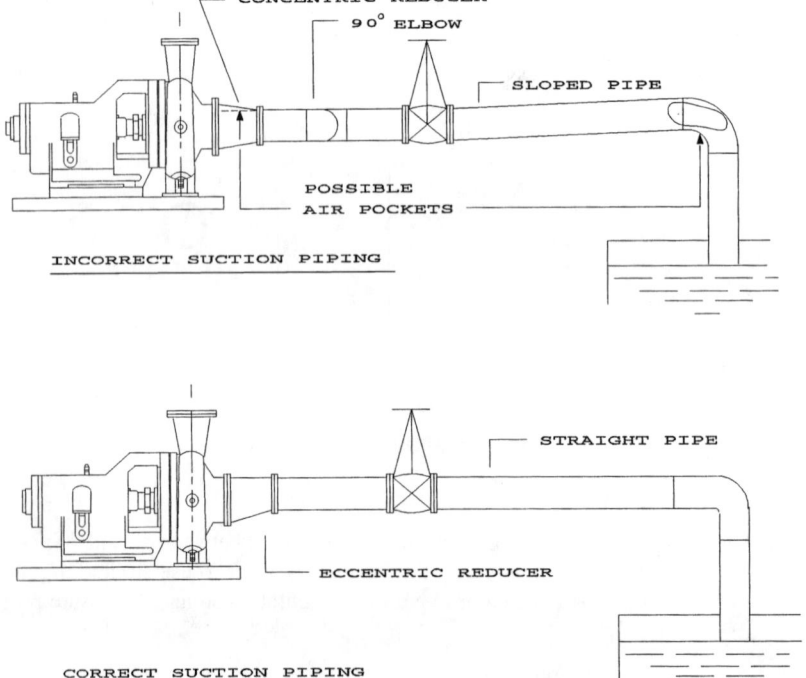

Figure 20.2 Undesirable and Correct Piping Configurations

Installation and Operation **149**

6. The discharge piping shall have a straight pipe with a length at least 2.5 times the discharge nozzle diameter.
7. Install a block valve in the suction pipe as close as possible to the pump.
8. Install a check valve and a block valve in the pump discharge piping. The block valve shall be downstream of the check valve.
9. The branch connection for a minimum flow bypass shall be upstream of the check valve.
10. Pumps with suction lift shall have a foot valve at the bottom of the suction line (Figure 20.3).

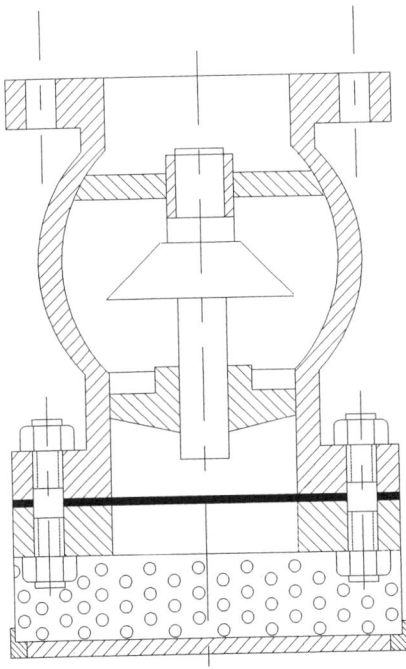

Figure 20.3 Foot Valve with Screen

Pump Start-up

Centrifugal Pumps

1. Before final shaft alignment, check pump and driver direction by bumping the driver a few times.
2. Before starting a centrifugal pump, fill the pump and the suction and discharge piping.

3. If the suction is flooded, open the pump suction block valve to fill the pump. If not successful, prime the pump and suction line. Use ejectors or vacuum pumps to prime the pumps.
4. While priming pumps, open the vent valve at the top of pump casing. Make sure no air or gas is trapped in the pump. Trapped air will cause the pump to cavitate.
5. Often the discharge piping is too long to fill by gravity. If the system has a minimum flow bypass, start the pump with the discharge block valve closed and the minimum flow bypass valve open. Then slowly open the discharge valve until full flow occurs.
6. With smaller pumps that do not have a bypass, slowly crack the discharge valve and keep opening it until the discharge line is full or until the pump pressure equals the system pressure.
7. If the pump has to pump against an already pressurized system, do not open the discharge block valve until the pump pressure equals the system pressure. Then open the valve slowly.
8. Do not operate the pump with the discharge or suction valve closed except for a very short time, because this will damage the pump.

Positive Displacement Pumps

1. Do not start or run a positive displacement pump with a closed discharge block valve.
2. Before starting a reciprocating pump, open the suction, discharge, vent, and bypass valves.
3. Turn the pump over manually to study ease of movement.
4. Run the pump at reduced speed for a few minutes before increasing to full speed for continuous operation.
5. Prime the pumps before starting them. Running dry usually damages rotary pumps. Therefore, make sure not to trap any air inside the pump.

Chapter 21

Troubleshooting

Centrifugal Pumps

These pumps may run years without problems, but some may evidence problems as soon as start-up. Problems that may occur, although the pumps performed well during testing, include the following items, each followed by the problem's possible causes:

No Flow

- Failure to prime the pump
- A loose impeller
- Faulty coupling

Insufficient Flow and/or Pressure

- System head higher than pump discharge pressure
- Insufficient NPSHA
- Suction strainer plugged or partially obstructed
- System head higher than pump discharge pressure
- Air and/or gas in media, usually through faulty or improperly installed gasket in the suction line
- Stuck foot valve
- Electric motor wired wrong, causing the pump to run in reverse
- Sharp edges in suction piping caused by undersized gaskets or burrs in pump casing
- Pump operating too far on the right or the left of the performance curve
- Suction lift higher than estimated
- Higher fluid viscosity than estimated
- Damaged impeller
- Suction piping not correct

Power Demand Increase

- Drop in pump efficiency. Many pump manufacturers increase pump efficiency by building pumps with tight running clearances. These clearances tend to open up after a while to become similar to that of their competitors' pumps, decreasing pump efficiency. As the efficiency drops, horsepower requirements increase. These will stabilize after a running-in period.

- As the wear rings wear out, the power demands will increase gradually. A slow increase in power demand indicates the wear rings need changing.
- Pump misalignment
- Change in liquid viscosity
- Impeller wear rings rubbing casing wear rings or pump bowls
- Damaged or worn bearings

Cavitation

- Insufficient NPSHA
- Air and/or gas is trapped in the pump or is entering the pump or suction piping. Leaking casing gaskets may cause this problem.
- Pump operates close to minimum flow, where the pump manufacturer may have stated NPSHR incorrectly.
- Pump operates too close to the end of the performance curve, where the NPSHA may not suffice.
- Clogged suction strainer. This is one of the major causes for pump malfunctioning. When examining pump problems, always look for clogged permanent strainers or forgotten temporary ones.
- Improperly designed pump pits. This accounts for many problems with cooling tower pumps. Often, these pumps are fairly large. Even though Engineering may follow the Hydraulic Institute's recommendation when designing the pit, these guidelines may not hold true for large flows. To avoid problems in these cases, have a specialized laboratory make flow simulations.

Damaged Bearings

- Pump shafts incorrectly aligned during installation, or misalignment occurring during operation
- Pump cavitation
- Bearing seal failure
- Bearing contamination from pump media
- Damaged pump bearing housing
- Improper bearing lubrication
- Bearings run hot because of insufficient bearing cooling
- Excessive suction pressure

Premature Seal Failure

- Damaged or worn bearings
- Sand or grit in pump media
- Improperly installed seals
- Pump misalignment
- Seal flush failure
- Rotating assembly imbalance

Reciprocating Pumps

Reciprocating pumps may also run years between failures, although they are higher maintenance machines than centrifugal and rotary positive displacement pumps. Therefore, Maintenance should stay alert for problems mentioned in this section.

Insufficient Capacity

- Slipping V-belts
- Air in cylinders
- Worn valves and seats
- Insufficient priming of all cylinders
- Insufficient suction pressure
- Damaged foot valve
- Clogged or partially clogged suction strainer

Excessive Pump Vibration

- Entrained air or gas in pump media
- Pump not sufficiently primed
- Pump running too fast
- Worn bearings
- Worn crosshead
- Broken or worn valve seats
- Piston rod is not tight enough
- Too much play in crosshead and/or crank pin

Premature Packing Failure

- Packing not installed correctly
- Pistons and/or rods improperly installed
- Improper packing lubrication
- Damaged plungers or rods

Excessive Mechanical Wear

- Sand or grit in pump medium
- Broken or damaged valve springs
- Air or gas entering the system
- Improper lubrication
- Water contaminating oil in crankcase
- Pump overloaded because of change in pumping conditions
- Pump media contaminating power end
- Improper material selection

Appendix 1

Sample Pump Specifications

NORD ENGINEERING INC. SPECIFICATION: 102	REV: A DATE: 02-10-97

Specification for End Suction Pump

Client: Sund Chemical Inc.

Auburn Plant

Project No.: 97012-01

Revision Record

Rev. No.	Description	By	Date	Approval
A	ISSUE FOR BID	UW	02-15-97	

End Suction Pump

TABLE OF CONTENTS

NO.	TITLE	PAGE
1.0	SCOPE	3
2.0	CODES, STANDARDS, SPECIFICATIONS, AND DRAWINGS	3
3.0	GENERAL	3
4.0	DESIGN AND FABRICATION	4
	4.1 END SUCTION PUMP	4
	4.2 ELECTRIC MOTOR	5
	4.3 SKID	5
5.0	INSTRUMENTATION AND CONTROL	5
6.0	PAINTING	6
7.0	INSPECTION AND TESTING	6
8.0	PREPARATION FOR SHIPMENT	6
9.0	WARRANTY	6
	ADDENDA: VENDOR DOCUMENT REQUIREMENTS	
	DATA SHEETS	

1.0 SCOPE

1.1 This specification describes the minimum requirements for the design, fabrication, assembly, inspection, testing, and painting of three (3) electric-motor driven end suction pumps, to be installed at Sund Chemical's Auburn plant.

1.2 In addition to the conditions and requirements of this specification, the pumps shall also be built in accordance with general specifications, electrical motor specification, drawings, and codes referenced herein.

1.3 Within this specification, the following definitions shall apply:

COMPANY: SUND CHEMICAL INC.
VENDOR: Successful bidder/manufacturer/supplier/contractor

2.0 CODES, STANDARDS, SPECIFICATIONS, AND DRAWINGS

2.1 Codes and Standards
VENDOR shall design, fabricate, and test the equipment in accordance with the following standards, agencies, and codes:

- *ANSI B73.1—American National Standards Institute, Specifications for Horizontal End Suction Centrifugal Pumps for Chemical Process*
- *Hydraulic Institute Complete Pump Standards*
- *NEMA—National Electrical Manufacturers Association MG 1, Motors and Generators*
- *OSHA—Occupational Safety and Health Act Standards*

3.0 GENERAL

3.1 The two end suction pumps shall be identical. They shall be suitable for duties shown on this specification and the attached data sheets.

3.2 The pumps shall be able to operate a minimum of 20,000 hr between maintenance and overhaul at the most exacting operating point.

3.3 The design conditions are:

- Liquid: Diesel fuel
- Flow rate: 220 gpm
- Specific gravity: 0.87
- Temperature (normal):   80°F
- Temperature (max.):   120°F
- Suction pressure: 20 psig
- Discharge pressure: 50 psig
- NPSHA   8 ft

3.4 The VENDOR shall design, purchase materials for, and fabricate the package as specified herein. The pumps shall be installed in a corrosive environment outdoors without shelter and shall be designed for this condition.

3.5 VENDOR shall supply pump performance curves with its proposal. The performance curves shall cover the range of operating conditions shown on the attached data sheets.

3.6 COMPANY approval of VENDOR drawings shall not relieve the VENDOR of the responsibilities to comply with all the paragraphs in this specification.

3.7 VENDOR shall supply all documents required by the attached Vendor Document Requirements in the correct quantities and on a timely schedule.

3.8 All burrs shall be removed from drilled bolt holes.

3.9 VENDOR shall provide stainless steel name tags and brackets. The name tags shall be according to *NEMA 37.1*.

4.0 DESIGN AND FABRICATION

4.1 *End Suction Pumps*

4.1.1 In case of conflict between this specification and the attached data sheets, the data sheets shall govern.

4.1.2 The pumps shall have a rising head capacity curve. The highest head shall be at pump shut-off. The pumps shall be able to deliver 120 percent rated capacity at no less than 60 percent rated discharge pressure.

4.1.3 The pumps shall operate in parallel and shall have the same head-rise to pump shut-off. The total shutoff head shall be at least 110 percent of rated pressure but shall not exceed 120 percent.

4.1.4 Rated capacity shall be less than capacity at the best efficiency point (BEP) but shall not be less than 70 percent of that capacity.

4.1.5 The pumps shall operate at 3,600 rpm. The pumps shall not be close coupled. The coupling between the pump and the electric motor driver shall be a flexible disc type with a spacer. VENDOR shall provide an OSHA-approved coupling guard. The coupling guard shall be nonsparking and removable. The guard shall be designed for a 200-lb concentrated load at mid-span.

4.1.6 Pumps that have a suction specific speed (N_S) of more than 11,000 require written approval by COMPANY.

4.1.7 The pumps shall have balanced mechanical seals, API Code BSTFM.

4.1.8 The impellers shall have replaceable front and back wear rings. These shall be press fitted or shrunk onto the impeller and secured by a lock pin. The wear ring mating surfaces shall have a different Brinell hardness.

4.1.9 The impeller diameters shall not exceed 95 percent of the maximum allowed impeller size.

4.1.10 Casings shall have vent and drain connections. The drains shall be flanged and piped to a common drain line that shall end with a flanged connection at the skid edge.

4.1.11 The shaft shall have replaceable protective sleeves in the stuffing box area. The sleeves shall be type 316 stainless steel.

4.1.12 The pumps shall have a gasket between the sleeves and a shaft shoulder to prevent leakage between the shaft and the sleeve. The sleeves shall be locked to the shaft either by a key or tongue-and-groove method.

4.1.13 Bearings shall be antifriction and shall have a minimum L-10 life of 25,000 hr at maximum thrust and radial loads. The thrust bearings shall be two back-to-back angular contact bearings.

4.1.14 Seal arrangements shall comply with *API Standard 610*, mechanical seal classification code.

158 *Practical Introduction to Pumping Technology*

4.1.15   All mechanical seal parts shall be type 316 stainless steel or better.
4.1.16   The seal piping shall be type 316 stainless steel with AISI 316 stainless steel tubing fittings. All parts of the flushing system shall be equal to or better than the pump casing material.
4.2   *Electric Motors*
4.2.1   The electric motor drivers shall be 480 v, 3 pH, 60 Hz squirrel-cage induction motors in a TEFC enclosure.
4.2.2   The electric motors, excluding service factor, shall be able to drive the pumps through the head capacity curves. The electric motors, minus the service factor, shall have a rating at least 110 percent of the BEP rating.
4.2.3   The motors shall have grease-lubricated journal bearings.
4.2.4   The couplings between the pumps and the motors shall be the flexible twin disc type with spacer.
4.2.4   The couplings shall be protected by OSHA-type removable coupling guards. The couplings shall be sized to handle full motor power with a minimum service factor of 1.75. Guards shall be designed to hold a concentrated load of 200 lb at mid-span.
4.3   *Skid*
4.3.1   VENDOR shall mount the equipment on a structural skid designed for a single-point lift from four lifting lugs placed symmetrically around the skid's center of gravity. The skid shall have raked ends.
4.3.2   Both main and cross-structural members shall be full depth and shall be joined web to web.
4.3.4   The drip and drain piping shall be 3 in. and shall slope three ways to both ends.
4.3.5   The VENDOR shall provide a $^{3}/_{16}$-in. checkered floor plate. The floor plate shall be flush with the runners, and both top and bottom shall be seal welded.
5.0   INSTRUMENTATION AND CONTROL
5.1   The VENDOR shall provide a local skid-mounted pneumatic NEMA 4X fiberglass control panel with stand, suitable for Class I, Division 2, Group D area classification, without air purge.
5.2   All internal braces, hardware, and instruments shall be type 316 stainless steel. The panel shall have a removable, full-opening door at the rear with heavy-duty type 316 stainless steel hinges.
5.3   The pump controls shall include a hands-off-auto switch (HOA) for each pump, as well as manual stop/start switches. The control panel shall also have a red "Run" pilot light for each pump. These devices shall be wired to terminals in the panel for interconnection to the separate motor starters.
5.4   The panel shall have a terminal strip for all termination to connect to the main plant control room.
5.5   The nameplates shall be laminated plastic, black with white letters, mounted with type 316 stainless steel machine screws with hex nuts.
5.6   COMPANY shall provide instrument air to the control panel. Air pressure shall be a minimum of 80 psig. Two pressure regulators shall reduce the air pressure to that required by the logic circuits.

6.0 PAINTING
The entire package shall be painted with COMPANY-approved three-coat marine paint.

7.0 INSPECTION AND TESTING

7.1 The COMPANY shall inspect the material and the workmanship on the package. The VENDOR shall replace, at its own cost, any material and workmanship not in compliance with this specification.

7.2 VENDOR shall not coat any pressure part or structural framing before completion of the last test.

7.3 VENDOR shall hydrotest all pressure parts to 1.5 times the maximum allowable working pressure (MAWP). There shall be no leakage during the 1½ hr test. The VENDOR shall use a COMPANY-approved corrosion inhibitor during the hydrotest.

7.4 The VENDOR shall run a performance test on one of the pumps. The test shall be according to the Hydraulic Institute.

7.5 Prior to shipment, the completed package shall be submitted to a witnessed test to confirm operation alignment and calibration of all components and instruments. The package shall be string tested at VENDOR's shop prior to shipment.

8.0 PREPARATION FOR SHIPMENT

8.1 After completing all tests, VENDOR shall cover all machined surfaces with a corrosion inhibitor. All loose items shall be placed in the same crating as the pump package. The loose items shall have metal identification tags.

8.2 A copy of the package assembly drawing in a waterproof envelope shall be stapled to the outside of the shipping crate. VENDOR shall send the COMPANY three (3) copies of the packing list and assembly drawings.

8.3 The shipping crate shall be clearly marked (use stencil) on each side and top as follows:

Sund Chemical No.:	97012-01
Sund Chemical Location:	Auburn, FL
P.O. No.:	0010-97
Item No.:	P-102, 103, 104

8.4 Male and female threaded correction shall be covered by a corrosion inhibitor and plugged with solid steel caps.

9.0 WARRANTY

9.1 The VENDOR shall have full responsibility for the design and performance of the pump package and shall guarantee materials and workmanship for 12 months after start-up or 18 months after shipment, whichever happens first.

9.2 Under this warranty, the VENDOR shall replace, at its own expense, all material and/or components that fail during the warranty period. The VENDOR shall also furnish, at its own expense, an experienced service person to supervise repairs and eventual replacements.

Appendix 2

Centrifugal Pump Data Sheet

NORD ENGINEERING INC. HOUSTON, TEXAS	CENTRIFUGAL PUMPS	No. A	Date 02/10/97	By UW	**REVISION** Issue For Bid	Job #: 97012–1 Spec #: 102.2
Client: Sund Chemical Inc Project: Auburn Plant Location: Florida Panhandle Service: Diesel Fuel Transfer Pumps				Equip No.	P-101, P-102, P-103	Date: 2/10/97 By: UW Rev: A Appv'd: KG

Mfg: * Type: Horizontal End Suction Pumps

OPERATING CONDITIONS—EACH PUMP | * PERFORMANCE

Liquid: Diesel Fuel	Operating Temperature: 80 °F	Proposal Curve No.:	
Rated Flow: 220 GPM	NPSH Available: * 8 Ft	NPSH Req'd (Water) Ft:	
Rated Disch Press.: 50 PSIG	Viscosity: 0.85 CS	No. of Stages:	RPM:
Specific Gravity: 0.87	Code: ANSI B73.1	Design Eff: %	BHP:
Vapor Pressure at PT: * PSIA		Max BHP Rated Impeller:	
Differential Pressure: 30 PSI	Erosion Caused by: No Erosion	Max Head Rated Impeller:	FT
Differential Head: 80 Ft	Corrosion Caused by: No Corrosion	Min Continuous Flow:	GPM
Design Temperature: 80 °F		Rotation Facing Coupling End:	

Water Cooling:
  Bearings:
  Stuffing Box:
  Pedestal:
  Gland:
Total Water Required GPM
Packing Cooling:
Flushing: API Plan:

CONSTRUCTION & MATERIALS

Case Mounting: Horizontal	Tapped Openings: Vent, Drain
Split: Radially	Radial Bearings: Antifriction
Type: Overhung	Thrust Bearings: Double Row
Base Plate: Fabricated	Angular Contact

Nozzles	Size	ANSI Rating	Facing	Position
Suction	3"	150 #	RF	End
Disch	2"	150 #	RF	Top

Impeller Diameter, Rated: * Max: * Type: Enclosed
Coupling & Guard: Mfgr: * Driver Half Mounted by: Vendor
( ) Packing: Mfgr & Type: Size: No. of Rings:
(X) Mech Seal: Mfgr & Model: API Class Code: BSTFM

AUX PIPING BY MFG

Cooling Water ( ) Tubing ( ) Pipe
Seal Flush (X) Tubing ( ) Pipe

SHOP TESTS REQUIRED WITNESSED

Factory
NPSH ( ) ( )

MATERIALS

Casing: Carbon Steel	Impeller: 316 SS
Shaft: 316 SS	Shaft Sleeves: 316 SS
Wear Ring Casing: 316 SS	Wear Ring Imp: 316 SS
Bearing Housing: Carbon Steel	Throat Bushing: 316 SS
Bearing H Cover: Carbon Steel	Gland Plate: Carbon Steel
Bolting, Studs: A-193-B7	Nuts: A-194-H4

Hydrostatic Test Pressure: PSIG
Max Allow Case Wp: PSIG @ °F
Weights, Pump: Base:
  Motor: Column: NA

MFG FINAL DATA (AS BUILT)

Outline Drawing No.:
Pump Sect Drawing No.:
Seal Diagram Drawing No.:
Wear Ring Clearance:
Test Curve Number:

MOTOR DRIVER

Mtd by: Pump Vendor Bearings: Antifriction
HP: * RPM: * Frame: * Full Load Amps: *
Mfr: * Area Classification: Class I, Div 2, Group D
Type: Induction Insul Class: F Space Heater: 120 V
Enclosure: TEFC Temp Rise: 40°C Volts/Phase/Hz: 460/3/60
Internally Epoxy Coat Motor (X) Antifungal Tropicalized Insulation ( )
Coupling Guard: Enclosed (X) Nonsparking (X)

OPTIONS

(X) Mechanical Seal ( ) Packing
(X) Installed ( ) Installed

NOTES: 1. * Denotes information to be provided by vendor with the quotation.
  2. Pump shall have a 316 SS nameplate permanently attached to the mounting plate.

Appendix 3

Internal Combustion Engine Data Sheet

NORD ENGINEERING INC HOUSTON, TEXAS	INT. COMB. ENGINE	No.	Date	By	REVISION	Job #: 8080 Spec #: 5800.2 Date: 11/29/88
Client: Sund Chemical Inc Project: Auburn Plant Location: Florida Panhandle						By: DAL Rev: A Appv'd: _____

Service: Pipeline Pump Driver   Equip No.: DE-102
Mfg & Model No.: *   Type: Diesel Engine

OPERATING DATA

Driven Equipment: Reciprocating Pumps   Synchronizing Requirements: ( ) Parallel ( ) Nonparallel
Approximate Load at Engine Shaft: 390 BHP   Turbocharging Acceptable: (X) Yes ( ) No
Duty: (X) Continuous ( ) Intermittent   Compression Ratio Preferred: ( ) Std (X) High
Preferred Speed: 1100 RPM, 90 % of Max Mfg Recom Engine Cooling: (X) Radiator ( ) Air Fin ( ) Shell & Tube
Type Drive Preferred: ( ) Direct (X) Gear ( ) V-Belt   Engine Cooling Furnished By: Vendor
Engine Starter: ( ) Pneumatic (X) Electric   Engine Speed Control: ( ) Manual (X) Automatic
Type of Start: (X) Manual ( ) Local ( ) Remote (X) Both   Engine Speed Control Signal: (X) Pneumatic ( ) Electric
Signal for Auto Start/Stop: (X) Pneumatic ( ) Electric

SITE CONDITIONS

Location: Florida Panhandle, Coastal Environment   Installation: (X) Land ( ) Platform ( ) Building
Altitude: 45 ft   Complete Engine Enclosure by Vendor: (X) Yes ( ) No
Ambient Temp: 85°F, Min: 25°F, Max: 105°F   Area Classification: ( ) Hazardous (X) Nonhazardous
Relative Humidity: 100%   Electrical Classification: Class: I, Div: 2, Group: D
Atmosphere: (X) Corrosive ( ) Noncorrosive

UTILITIES AVAILABLE

Electrical Power: 480 Volts, 3 Phase, 60 Hz   Fuel Type: ( ) Natural Gas (X) Diesel
Instrument Air: 80 PSIG Min   Fuel Gas Quality: ( ) Clean & Dry ( ) Wet & Dirty
Starting Gas: (X) Air ( ) Nat Gas   Fuel Gas Press: PSIG: LHV: BTU
Cooling Water: (X) Freshwater ( ) Seawater
Cooling Water Conditions: As Required GPM at 85°F

ENGINE PERFORMANCE DATA (BY VENDOR)

Engine Mfg/Model No.:   Basis for Determining BHP:
Engine Type: Rating Temperature: 80°F   Elevation: 45 Ft
Rated Horsepower: BHP Max at: RPM   Engine Derating % Bare Engine: %
Cycle: No. of Cylinders: BMEP at Rated HP (This Application):
Cylinder Bore: Stroke: In   Deration for Location Elev and Temp: BHP
Compression Ratio: Displacement: In^3 Less Horsepower for Engine Accessories: BHP
Recommended RPM: Max: Min: Available Horsepower: BHP
Engine Heat Rejection: BTU/HR   Fuel Gas Pressure Required   PSIG

CONTROLS AND COMPONENTS (BY VENDOR)

Radiator Mfg/Model No.:   Gear Reducer:
Fan Mfg/Model No.:   Gear Ratio:
Oil Filter: AGMA Rating:
Oil Cooler: Engine Instrument Panel:
Air Filter: Tachometer:
Air Inlet Piping Included: Hourmeter:
Fuel Pump: Jacket Water Temp Gauge:
Water Pump: Lube Oil Temp Gauge:
Ignition System: NA   Lube Oil Pressure Gauge:
Alternator: Fuel Gas Pressure Gauge:
Lube Oil Heater:
Jacket Water Heater: Shutdown Devices:
Governor: Low Oil Pressure:
Vibration Isolators: High Cooling Water Temp:
Prelube Pump (If Required):   High Oil Temp:
Muffler (Include DB Rating):   Overspeed:
Coupling: Vibration:
Starter:

NOTES: 1. Engine(s) shall have 316 SS nameplate permanently attached, engraved with tag number.

Appendix 4

Electric Motor Data Sheet

NORD ENGINEERING INC Houston, Texas	ELECTRIC MOTORS	No.	Date	By	REVISION	Job #: 97012–02 Spec #: 102
Client: Sand Chemical Inc Project: Auburn Plant Location: Florida Panhandle		A	02/10/97	UW	Issued For Bid	Date: 2/15/97 By: UW Rev: A Appv'd: KG

Service: Diesel Transfer Pump Driver   Equip No. PM-101, PM-102, PM-103

Mfg: *   Type: Squirrel-Cage Induction Motor

MOTOR DESIGN DATA | * MANUFACTURER'S DATA

Basic Data
* kW   *   RPM   *   SF   *
460 V, 3 Phase, 60 Hertz
Ambient Temperature: 80°F   Altitude: 45 Ft

Type Motor: Induction
Synchronous @ 1.0 PF: 0.8 PF Leading:
Brushless Synchronous:

Enclosure

Explosion Proof ( )   TEFC (Totally Enclosed Fan Cooled) (X)
Weather Protected Type II, with Filters & Screens ( )
Drip-Proof For Indoor Use ( )
With Tropical Protection ( )
Other (X)   See Note 3

Starting

Full Voltage (X)   Reduced Voltage ( )   Percentage Approx:
Loaded (X)   Percent Loaded:

Frame: Full Load RPM:
Efficiency,   FL: 3/4L: 1/2L:
Power Factor, FL: 3/4L: 1/2L:
Current, Full Load: Amps
         Locked Rotor: Amps
Locked Rotor, Max Allow Time: s Hot
                                s Cold
Power Factor: at 100% Voltage
Acceleration Time: at 100% Voltage
Temperature Rise: °C by:
     Total Allowable: °C
Insulation, Class: Type:
No. Consecutive Starts   in   Min
Torque: Lbs-Ft Full Load
     Starting: %L

Rotation Facing Coupling End:
Motor WR Sq: Lbs-Sq Ft
Reactances, X'D: X'D:
Bearings, Type: Lubr:
Total End Float: Limit to:

ACCESSORY EQUIPMENT | SYNCHRONOUS MOTOR FIELD DATA

(X) Base Plate   ( ) Sole Plate
(X) Terminal Box Furnished with:
   (X) Surge Protection (Arrestors & Capacitors)
   (X) Grounding Terminal
   ( ) Mounting only of 3-Differential CT

( ) Leads Arranged for Differential Protection
   ( ) RTDs, NUMBER: MATERIAL:
   ( ) Rating: Ohm at   °F
(X) Space Heaters 220/120 Volts   Watts
                   1 Phase   60 Hertz
( ) Bearing Temperature Relays   ( ) Explosion Proof
( ) Synchronous Motor DC Excitation
   ( ) Furnish Static Supply System
( ) Purged
( ) Mount Half Coupling (Furnished by Others)
( ) Other:

Excitation: V,   KW
Current: AMP at Full Load

Resistance: _____ OHM at 75°C
Allow Stall Time & Induced Field
Current at Rated Voltage:
% Speed: 0   50   75   95
  Seconds:
  Amps:
Field Disch Resistor: OHM
Full Voltage Line Current @ 95%
Speed:
Max Line Current, First Slip Cycle @
Pull Out:

BRUSHLESS EXCITER FIELD DATA

Excitation: V DC @ Exciter Field Term
Current: A DC @ Full Load
            A DC @ No Load
Discharge Resistor: OHM

APPLICABLE SPECIFICATIONS | WEIGHTS AND DIMENSIONS

1. NEFPA 70, National Electric Code
2. NEMA
3. * Denotes information that shall be provided by Vendor

Net Weight: *   Lbs
Shipping Weight: *   Lbs
Max Erection Weight: * ___   Lbs
* _____ L * _____ W * _____ H

Appendix 5

Centrifugal Pump Package

Vendor Document Requirements

Documents	Required With Proposal	Required for Approval	Certified
Dimension drawing	4	8	8
Welding procedures		8	
Weights, dry/operating		8	8
Completed data sheets		8	8
Performance data curves	4	8	8
Panel layout		8	8
Hydrotest reports			8
Bill of materials		8	8
Recommended spare parts			8
Operating manual			8
Maintenance manual			8
Lubrication requirement			

Notes:
1. The final drawings shall be full size and shall include one reproducible mylar.
2. All approval drawings shall include one sepia.

Appendix 6

Maximum Viable Suction Lifts at Various Altitudes

Altitude	Barometric Pressure		Head of Water (ft)	Maximum Suction Lift (ft)
	psia	in. of Hg		
Sea level	14.7	29.9	34.0	21.3
1,000 ft	14.2	28.9	32.8	20.5
2,000 ft	13.7	27.8	31.6	19.3
3,000 ft	13.2	26.8	30.5	18.1
4,000 ft	12.7	25.8	29.3	17.2
5,000 ft	12.2	24.9	28.1	16.1
6,000 ft	11.8	23.9	27.3	15.4
7,000 ft	11.3	23.1	26.1	14.2
8,000 ft	10.9	22.2	25.2	13.7
9,000 ft	10.5	21.4	24.2	12.5
10,000 ft	10.1	20.6	23.3	11.7

Appendix 7

Suggested List of Vendors

COUPLINGS

Ameridrives International, Coupling Products
1802 Pittsburgh Ave., P.O. Box 13801, Erie PA 16502-1551
Phone: 814-480-5100

Lovejoy, Inc.
2655 N. Wisc. Ave., Downers Grove IL 60515-4299
Phone: 630-852-0500 (Fax: 630-852-2120)

Rex Nord Corp.
4701 W. Greenfield Ave., Milwaukee WI 53214-5300
Phone: 414-643-3000 (Fax: 414-643-3078)

ELECTRIC MOTORS

ABB Industrial Systems, Inc.
16250 W. Glendale Dr., New Berlin WI 53151-2840
Phone: 414-785-3200 (Fax: 414-785-3290)

Baldor Electric Co.
5711 S. Boreham Jr. St., Ft. Smtih AR 72903-0000
Phone: 501-646-4711 (Fax: 501-648-5792)

GE Motors Production
1635 Broadway, Ft. Wayne IN 46802
Phone: 219-428-2000

GEC Alstohm International
4-T Skyline Dr., Hawthorne NY 10532
Phone: 914-345-5100 (Fax: 914-347-5432)

Marathon Electric Mfg. Corp.
P.O. Box 08003, Wausau WI 54402-8003
Phone: 715-675-3311 (Fax: 715-675-6391)

166 *Practical Introduction to Pumping Technology*

Reliance Electric
24701 Euclid Ave., Cleveland OH 44117-1794
Phone: 216-383-6600 (Fax: 216-383-6036)

Siemens Energy & Automation Inc., Motors and Drives Division
4620 Forest Ave., Cincinnati OH 45212-3306
Phone: 513-841-3100 (Fax: 513-841-3290)

Toshiba International Corp., Industrial Division
13131-T W. Little York Rd., Houston TX 77041-5807
Phone: 713-466-0277

U.S. Electrical Motors Corp.
23335 Lasalle Lane, Sherwood OH 97140
Phone: 503-625-8994

Westinghouse Motor Co.
7300 W. Tidwell Rd., Houston TX 77040
Phone: 713-939-8868

GEARS

General Purpose

Boston Gear
14 Hayward St., Quincy MA 02171-2418
Phone: 617-328-3300

The Cincinnati Gear Co.
4400 Woodster Pike, Cincinnati OH 45227
Phone: 513-271-7700 (Fax: 513-271-0049)

Cleveland Gear Co.
3249 E. 80th St., Dept. TR, Cleveland OH 44104
Phone: 216-641-9000

Falk Corp., Subsidiary of Sundstrand
P.O. Box 492, Dept. TR, Milwaukee WI 54201
Phone: 414-937-4284 (Fax: 414-937-4359)

Lufkin Industries
P.O. Box 849, Dept. K, Lufkin TX 75902
Phone; 409-637-5738 (Fax: 409-637-5774)

Philadelphia Gear Corp.
181-T Gulph Rd., King of Prussia PA 19406
Phone: 610-265-3000 (Fax: 610-337-5637)

Right Angle Gears

Amarillo Gear Co.
P.O. Box 1789, Amarillo TX 79105
Phone: 806-622-1273 (Fax: 806-622-3258)

INTERNAL COMBUSTION ENGINES

Caterpillar
100 NE Adams, Peoria IL 61629
Phone: 309-675-1000

Deer Power Systems
P.O. Box 5100, Waterloo IA 50704
Phone: 319-292-6060 (Fax: 319-292-5075)

Detroit Diesel Corp.
13400 W. Outer Dr., Detroit MI
Phone: 313-592-5000

Dresser Industries, Waukesha Engine Division
1000-T W. St. Paul Ave., St. Paul WI 53188
Phone: 414-547-3311 (Fax: 414-549-2795)

Waukesha-Pearce Industries, Inc.
12320 S. Main St., Houston TX 77035
Phone: 713-723-1050 (Fax: 713-551-0454)

PUMPS

Centrifugals

American Machine & Tool Co.
400 Spring St., Rogersford PA 19468
Phone: 610-948-3800 (Fax: 610-948-5300)

Aurora Pump
800-T Airport Rd., North Aurora IL 60542
Phone: 630-859-7000 (Fax: 630-859-7050)

Crane Deming Pumps
1453 Allen Road, Salem OH 44460
Phone: 216-337-7861 (Fax: 216-337-8122)

Crane Pumps & Systems Inc., Weinman Division
420 Third St., Piqua OH 45356
Phone: 937-773-2442 (Fax: 937-773-2238)

David Brown Pumps, Inc.
7322 SW Fwy., Houston TX 77074
Phone: 713-981-3836 (Fax: 713-776-2442)

The Duriron Co. Inc., Pump Division
P.O. Box 8820, Dayton OH 45402
Phone: 513-226-4000 (Fax: 513-226-8122)

Fairbanks Morse Pump Co.
3601 Fairbanks Ave., Kansas City KS 66110-2918
Phone: 913-371-5000 (Fax: 913-371-2272)

Goulds Pumps
240 Fall St., Seneca Falls NY 13148
Phone: 315-568-2811 (Fax: 315-568-7709)

Ingersoll-Dresser Pump Co.
150-T Allen Rd., Liberty Corner NJ 07938
Phone: 908-647-6800 (Fax: 908-604-8195)

ITT A-C/ITT Marlow
1150 Tennesse Ave., Cincinnati OH 45229
Phone: 513-482-2500 (Fax: 513-482-2569)

Mission—Fluid King (National Oilwell)
P.O. Box 2108, Houston TX 77056
Phone: 713-462-4110 (Fax: 713-462-3152)

Paco Pump, Inc.
800-T Koomey Rd., Brookshire TX 77423
Phone: 800-955-5847 (Fax: 713-934-6082)

Peerless Pumps (Sterling Fluid Systems)
P.O. Box 7026, Indianapolis IN 46202
Phone: 317-925-9661 (Fax: 317-924-7202)

Price Pump Co.
P.O. Box Q, Sonoma CA 95476
Phone: 707-938-8441 (Fax: 707-938-0764)

Sulzer Bingham Pumps Inc.
2800 N.W. Front Ave., Portland OR 97210
Phone: 503-226-5200 (Fax: 503-226-5460)

Sundstrand Fluid Handling
14845 64th Ave., Arvada CA 80007
Phone: 303-452-0800 (Fax: 303-452-0896)

Metering Pumps

Plunger Type
American Lewa Inc.
132 Hopping Brook Rd., Holliston MA 01746
Phone: 888-539-2123 (Fax: 508-429-8615)

Jaeco-Stewart
23 Francis J. Clarke Circle, Ste. 2, Bethel CT 06801
Phone: 203-743-3703 (Fax: 203-743-4362)

Milton Roy, Flow Control Division
201 Ivyland Rd., Ivyland PA 18974
Phone: 215-441-7821 (Fax: 215-441-8621)

Diaphragm Type
American Lewa Inc.
132 Hopping Brook Rd., Holliston MA 01746
Phone: 888-539-2123 (Fax: 508-429-8615)

The Duriron Co., Inc., Pump Division
P.O. Box 8820, Dayton OH 45402
Phone: 513-226-4000 (Fax: 513-226-8122)

Jaeco-Stewart
23 Francis J. Clarke Circle, Ste. 2, Bethel CT 06801
Phone: 203-743-3703 (Fax: 203-743-4362)

Milton Roy, Flow Control Division
201 Ivyland Rd., Ivyland PA 18974
Phone: 215-441-0800 (Fax: 215-441-8621)

Pulsafeeder, a unit of IDEX Corporation
2883 Brighton Henrietta Town Line Rd., Rochester NY 14623
Phone: 716-292-8000 (Fax: 716-424-5619)

Warren Rupp, a unit of Idex Corp.
P.O. Box 1568, Mansfield OH 44901
Phone: 419-524-8388 (Fax: 419-522-7867)

Wilden Pumps & Engineering Co.
22069 van Buren St., Grand Terrace CA 92313
Phone: 909-422-1700 (Fax 909-783-3440)

Fluid Powered
Williams Instrument Co.
25217-T Rye Canyon Rd., Santa Clarita CA 91355
Phone: 805-257-2250 (Fax: 805-257-7963)

Vertical Turbine and Can Pumps

Goulds Pumps
240 Fall St., Seneca Falls NY 13148
Phone: 315-568-2811 (Fax: 315-568-7709)

Ingersoll-Dresser Pump Co.
150-T Allen Rd., Liberty Corner NJ 07938
Phone: 908-647-6800 (Fax: 908-604-8195)

Johnston Pump Co.
800-T Koomey Rd., Brookshire TX 77423-8803
Phone: 713-934-6009 (Fax: 713-934-6090)

Peerless Pumps (Sterling Fluid Systems)
P.O. Box 7026, Indianapolis IN 46202
Phone: 317-925-9661 (Fax: 317-924-7202)

Sulzer Bingham Pumps Inc.
2800 N.W. Front Ave., Portland OR 97210
Phone: 503-226-5200 (Fax: 503-226-5460)

Submersibles

Goulds Pumps
240 Fall St., Seneca Falls NY 13148
Phone: 315-568-2811 (Fax: 315-568-7709)

Ingersoll-Dresser Pump Co.
150-T Allen Rd., Liberty Corner NJ 07938
Phone: 908-647-6800 (Fax: 908-604-8195)

Peerless Pumps (Sterling Fluid Systems)
P.O. Box 7026, Indianapolis IN 46202
Phone: 317-925-9661 (Fax: 317-924-7202)

Sulzer Bingham Pumps Inc.
2800 N.W. Front Ave., Portland OR 97210
Phone: 503-226-5200 (Fax 503-226-5460)

Submersible Contract Pumps

ITT Flygt Corp.
35 Nutmeg Dr., Trumbull CT 06611
Phone: 203-380-4700 (Fax: 203-380-4705)

Little Giant Pump Co.
P.O. Box 12010, Oklahoma City OK 73157-2010
Phone: 405-947-2511 (Fax: 405-942-2285)

Pumpex, Inc.
103A Molasses Hill Rd., Lebanon NJ 08833
Phone: 908-730-7004 (Fax: 908-739-7580)

Sump Pumps

Crane Deming Pumps
1453 Allen Road, Salem OH 44460
Phone: 216-337-7861

Goulds Pumps
240 Fall St., Seneca Falls NY 13148
Phone: 315-568-2811 (Fax: 315-568-7709)

Ingersoll-Dresser Pump Co.
150-T Allen Rd., Liberty Corner NJ 07938
Phone: 908-647-6800 (Fax: 908-604-8195)

Peerless Pumps
P.O. Box 7026, Indianapolis IN 46202
Phone: 317-925-9661 (Fax: 317-924-7202)

Rotary Pumps

Sliding Vane
Blackmer Pumps
1809 Century Ave. S.W., Grand Rapids MI 49509
Phone: 616-241-1611 (Fax: 616-241-3752)

Rotary Gear
Ingersoll-Dresser Pump Co.
150-T Allen Rd., Liberty Corner NJ 07938
Phone: 908-647-6800 (Fax: 908-604-8195)

Pulsafeeder (IDEX)
2883 Brighton Henrietta Town Line Rd., Rochester NY 14623
Phone: 716-292-8000 (Fax: 716-424-5619)

Roper Pump Co.
P.O. Box 269, Commerce GA 30529
Phone: 706-335-5551 (Fax: 706-335-5505)

Tuthill Corp., Pump Division
12500 S. Pulaski Rd., Aslip IL 60803
Phone: 708-389-2500 (Fax: 508-388-0869)

Viking Pump Inc. (IDEX)
406 State St., Cedar Falls IA 50613-0008
Phone: 319-266-1741 (Fax: 319-273-8157)

Screw Pumps
IMO Pump Co.
P.O. Box 5020, Monroe NC 28111-5020
Phone: 704-289-6511 (Fax: 704-289-9273)

Shanley Pumps, Inc.
2525 S. Clearbrook Dr., Dept. LS, Arlington Heights IL 60005
Phone: 847-439-9200 (Fax: 847-439-9388)

Progressive Cavity Pumps
Moyno Industrial Products (Robbins & Meyer)
P.O. Box 960, Springfield OH 45501
Phone: 513-327-3553 (Fax: 513-327-3572)

Roper Pump Co.
P.O. Box 269, Commerce GA 30529
Phone: 706-335-5551 (Fax: 706-335-5505)

Solar-Powered Pumps

Shurflow
12650 Westminster Ave., Santa Ana CA 92706
Phone: 800-394-7709 (Fax: 714-554-4721)

Reciprocating Pumps

CAT Pumps
1681 94th Ln. N.E., Minneapolis MN 55449
Phone: 612-780-5440 (Fax: 612-780-2958)

David Brown Pumps, Inc.
7322 SW Fwy. Houston TX 77074
Phone: 713-981-3836 (Fax: 713-776-2442)

Gardner-Denver Ajax OPI Pumps
1800 Garner Expressway, Quincy IL 62301
Phone: 217-222-5400 (Fax: 217-224-7814)

Ingersoll-Dresser Pump Co.
150-T Allen Rd., Liberty Corner NJ 07938
Phone: 908-647-6800 (Fax: 908-604-8195)

Mission—Fluid King (National Oilwell)
P.O. Box 2108, Houston TX 77056
Phone: 713-462-4110 (Fax: 713-462-3152)

Wheatley-Gaso, Inc.
P.O. Box 3249, Tulsa OK 74101
Phone: 918-447-4600 (Fax: 918-447-4677)

Archimedes Screw Pumps

U.S. Filter/CPC
P.O. Box 36, Sturbridge MA 01566
Phone: 508-447-7344 (Fax: 508-347-7049)

TURBINES

Gas Turbines

ABB Turbine Power Division
1460 Livingston Ave., North Brunswick NJ 08902
Phone: 732-932-6000 (Fax: 732-932-6194)

G-E Co.
3135 Easton Tpke., Fairfield CT 06431
Phone: 800-626-2004 (Fax: 518-869-2828)

Steam Turbines

ABB Turbine Power Division
1460 Livingstone Ave., North Brunswick NJ 08902
Phone: 732-932-6000 (Fax: 732-932-6194)

Coppus Murray Group—Turbine Division
P.O. Box 8000, Millbury MA 01527-8000
Phone: 508-756-8391 (Fax: 508-756-8375)

Elliot Co.
901 N. Fourth St., Jeanette PA 15644-0800
Phone: 800-488-4242 (Fax: 724-527-8442)

G-E Co.
3135 Easton Tpke., Fairfield CT 06431
Phone: 800-626-2004 (Fax: 518-869-2828)

Siemens Power Corp.
1301 Ave. of the Americas, New York NY 10001
Phone: 212-258-4920

Pignoni Inc.
10000 Richmond, Houston TX 77042
Phone: 713-952-8374

Rollys Royce, Cooper Energy Services
105 No. Sandusky St., Mt. Vernon OH 43050-2495
Phone: 740-393-8200 (Fax: 740-393-8373)

Solar Turbines
P.O. Box 85376, San Diego CA 92186
Phone: 619-544-5000 (Fax: 619-544-2683)

Appendix 8

API-610 Mechanical Seal Classification Code

First Letter: B = balanced
U = unbalanced
Second Letter: S = single
T = tandem
D = double
Third Letter: End Plate Type
P = plain
T = throttle bushing
A = auxiliary sealing device

Fourth Letter:

	E	F	G	H	I	R	X
Static Seal Ring Gasket							
Seal Ring to Sleeve Gasket	Viton	Viton	TFE	Nitrile-Buna N	FKM Elast.	Graphite Foil	As Specified

Fifth Letter:

	J	K	L	M	N	X
Seal Ring	Carbon	Carbon	Carbon	Carbon	Carbon	As Specified
Mating Seal Ring	Stellite	Ni-Resist	Tungsten Carbide Co-Binder	Tungsten Carbide Ni-Binder	Silicon Carbide	As Specified

References

Baumeister, Theodore. *Mark's Standard Handbook for Mechanical Engineers,* 10th ed. New York: McGraw-Hill, 1996.

Benaroya, Alfred. *Fundamentals and Application of Centrifugal Pumps for the Practicing Engineer.* Tulsa, Okla.: Petroleum Publishing, 1978.

Florjancic, Dusan. *Net Positive Suction Head for Boiler Feed Pumps.* Winterhur, Switzerland: Sulzer Brothers, 1980.

_____. *Pumps for Fluid Transport: General Design Concepts.* Winterhur, Switzerland: Sulzer Brothers, 1979.

Frick, Thomas C. *Petroleum Production Handbook.* Dallas: Society of Petroleum Engineers of AIME, 1962.

Fritsch, Horst. *Metering Pumps.* Landsburg/Lech, Germany: Verlag Moderne Industrie AG: 1990.

Hydraulic Institute Complete Pump Standards, 4th ed. Cleveland: Hydraulic Institute, 1994.

The Hydraulic Institute Engineering Data Book, 2nd ed. Cleveland: Hydraulic Institute, 1990.

Karassik, Igor J. *Centrifugal Pump Clinic.* New York: Marcel Dekker, 1971.

_____, et al. *Pump Handbook,* 2nd ed. New York: McGraw-Hill, 1986.

Kirk, Franklin W., and Nicholas R. Rimboi. *Instrumentation,* 3rd ed. Homewood, Ill.: American Technical Publishers, 1975.

Mayer, Ehrhard. *Burgmann Mechanical Seals,* 3rd ed. Boston: Butterworth Scientific, 1982.

Pini, G., and J. Weber. *Material for Pumping Seawater and Media With High Chloride Content.* Winterhur, Switzerland: Sulzer Brothers, 1979.

Redmon, James D. *Selecting Second Generation Duplex Stainless Steel.* New York: McGraw-Hill, 1986.

Vlamming, D.J. *A Method for Estimating the Net Positive Suction Head Required by Centrifugal Pumps.* New York: American Society of Mechanical Engineers, 1981.

Warring, R.H. *Pumping Manual,* 7th ed. Houston: Gulf Publishing, 1984.

_____. *Pumps: Selection, Systems, and Applications,* 2nd ed. Houston: Gulf Publishing, 1984.

Yedidiah, S. *Centrifugal Pump Problems.* Tulsa, Okla.: Petroleum Publishing, 1980.

Index

A

Absolute pressure, 2
Absolute viscosity, 55
Acceleration, 61
Affinity formulas, 17–18
Alternate-current (A-C) motors
    full-load speeds, 101
    voltage and horsepower ratings, 100
Amplitude, 61, 62
Angular contact bearings, 86
Annunciators, 137–138
Antifriction bearings, 84–85, 104
Archimedes screw pumps, 40, 100
    diagram, 40
Atmospheric pressure, 1
Atmospheric tank, 68
    diagram, 67
Austenitic stainless steel, 97–98
Axial-flow pumps, 22
Axially split pumps, 27
    diagram, 26
    multistaged volute pump
        diagram, 28, 29

B

Bearings
    angular contact, 86
    antifriction, 84–85, 104
    cooling, 91
    damage, 152
    electric motor, 104
    inboard, 83
    internal product-lubricated, 83
    journal, 83, 104
    line-shaft, 89
    lubrication, 89–91
    outboard, 83
    roller, 87, 104
    seals, 91
    sleeve, 83, 104
    thrust, 88, 104
    tilted-pad, 87
    vertical pump, 88–89
Between bearing pumps, 27
    diagram, 26
Block valves, 134–135
Brake horsepower, 7, 15–16
    curves, 45
Buffer fluid schemes, 82

C

Can pumps, 32
Cantilever pumps, 25
Capacity control, 129–131
Cast iron, 93
Cavitation, 7, 152
Centrifugal pumps, 21–30
    curves, 45–54
    data sheet, 160
    inspection requirements, 142
    package, 163
    start-up, 149–150
    troubleshooting, 151–152
Closed loop system, 129
Control systems, 129
Control valve types, 128–129
Conversion factors, 1, 4, 15, 21

177

Corrosion, 92–93
Corrosive liquids (recommended material), 95, 96
Couplings
  diaphragm, 124
  disc, 125
  flexible, 121–123
  gear, 124
  hydraulic (variable speed), 131
  internal combustion engine, 108
  jaw, 123
  pin-and-bushing, 123
  rigid, 121
  roller chain, 124
  sleeve, 123
Critical speed, 7, 61
Crude oil tank with nitrogen blanket, 70–71
Cyclone separator, 82

D

Data sheets, 42–43, 141, 160–162
Density, 5
Dial thermometers, 137
Diaphragm couplings, 124
Diaphragm pumps, 31, 33–35
  diagram, 34
Disc couplings, 125
Discharge control valves, 129–130
Discharge pressure, 5, 10
Displacement, 7, 61
Documentation, 140
Double mechanical seals, 80, 81
Duplex stainless steel, 98
Duplex stainless steel compositions (table), 95
Dynamic viscosity, 55

E

Eddy current principle, 131
Efficiency curves, 45
Elastomer linings, 99
Electric motors, 100–106
  acceleration, 105
  bearings, 104
  data sheet, 161
  efficiency, 105
  insulation, 104–105
  motor enclosures, 103
  phases, 103
  variable-speed control, 131
Electrical area classification, 139
End suction pumps, 22–23
  diagrams, 23, 24
Engines, internal combustion, 106–109
Engler viscosity meter, 56
Epicyclic gears, 120

F

Feed forward system, 129
Feedback system, 129
Ferritic steel, 93, 97
Filter in (filtered), 62
Filter out (unfiltered), 62
Flexible couplings, 121–123
Flow, 3
  minimum, 7
Flush and quench fluids, 82
Frequency, 62
Friction, 9
  loss calculation, 9

G

Gas turbines, 111–112, 131
Gauge pressure, 2
Gear contact pattern, 144
Gear couplings, 124
Gear pumps, 35
  diagrams, 36, 37
Gears
  epicyclic, 120
  helical, 115–117
  parallel shaft, 114–117
  right-angle, 118–120
  spur, 114, 115
Grease lubrication (bearings), 89–90

Index **179**

H

Hazardous locations, 102–103
Head
 calculations, 10–15
 discharge, 5, 12, 13, 15
 loss calculations, 9
 net positive suction (NPSH), 5
 suction, 5, 13, 14
 total differential, 5
 velocity, 6–7, 10
Head capacity (H-C) curves, 17, 22, 45–48, 49, 54, 71
Helical gears, 115–117
Horizontal split-case pumps, 27
 diagram, 26
Horsepower, 7, 15–16
Hydraulic couplings (variable speed), 131
Hydraulic drives, 113, 131
Hydraulic horsepower, 7, 15–16
Hydraulic recess area, 77
Hydrostatic test, 143

I

Impellers, centrifugal pumps, 22
Inboard bearing, 83
Induction motors, 100
In-line pumps, 23
 diagram, 24
Inspection, 142–143
Interface area, 77
Internal combustion engines, 106–109
 barring device, 109
 cooling system, 108
 coupling, 108
 data sheet, 161
 engine speed, 107–108
 engine types, 106–107
 exhaust, 108
 governor, 109
 instrumentation, 109
 power ratings, 107
 starting system, 108
 variable-speed control, 131

Internal product-lubricated bearings, 83

J

Jaw couplings, 123
Joints, universal, 125
Journal bearing, 83, 104

K

Kinematic viscosity, 55

L

Line-shaft bearings, 89
Line-shaft pumps, 29–30
 diagram, 30
Linings, elastomer, 99
Liquid level control, 132
Lobe pumps, 39
 diagram, 39

M

Martensitic stainless steel, 97
Mechanical face seals, 74, 75–82
 balanced and unbalanced, 77–78
 classification code, 175
 diagrams, 76, 77, 78, 79, 81
 face material, 80–82
 general design, 75–76
 life expectancy, 82
 seal arrangements, 77–78
 seal types, 78–80
Mechanical run test, 144
Mechanical seal leak detection, 138, 139
Metering pumps, 31
Metric conversions, 1
Minimum flow bypass, 8, 132
Mixed-flow pumps, 22
Modulating control, 133
Modulating valves, 128–129
Motors
 alternate-current (A-C), 100–101

180   *Practical Introduction to Pumping Technology*

electric, 100–106
induction, 100

N

Net positive suction head
    (NPSH), 5, 66–73
  calculations, 66–71
  curves, 45
  definition, 66
  test, 145

O

Oil lubrication (bearings), 90–91
On-off control, 133
On-off valves, 128
Open loop system, 129
Operating point, 48
Outboard bearing, 83
Overheating, 138
Overhung pumps, 22–23

P

Packed glands, 75
Parallel shaft gears, 114–117
Passive layer, 92
Performance curve checklist, 141
Performance test, 143
Peristaltic pumps, 41
  diagram, 41
Phase, 62
Pin-and-bushing couplings, 123
Piston pumps, 31
  diagram, 33
Pitot tube pumps, 41
Plastic, 99
Plunger pumps, 31
  diagram, 32
Positive displacement pumps, 30–41
  capacity control, 134
  control selection, 134–136
  curves, 54
  start-up, 150
  stroke adjustment, 136
  variable-speed control, 135

Power, 101
  solar, 113
Power pumps, 31
Pressure, 1
  absolute, 2
  atmospheric, 1
  control, 133–134
  discharge, 5, 10
  gauge, 2
  suction, 6
  vapor, 2
  vessel, 69–70
Pressure relief valves, 133–134
Progressive cavity pumps, 38
  diagram, 38
Proximity probe, 62
Pulsation dampeners (bottles), 135, 136
Pump
  calculations, 9–18
  installation, 146–147
  materials, 93–99
  specifications, 19–20, 42–44
  start-up, 149–150
  types, 21–41
Pumps in parallel, 48–52
Pumps in series, 51–54

Q

Quench and flush fluids, 82

R

Radial-flow pumps, 22
Reciprocating pumps, 30–35
Redwood viscosity meter, 56
Relief valves, 133
Resistance temperature detectors
    (RTD), 137
Resonance, 62
Revision numbering sequences, 43
Right-angle gears, 118–120
Rigid couplings, 121
RMS level, 62
Roller bearings, 87, 104
Roller chain couplings, 124
Rotary pumps, 35

S

Safety checklist for piping and valves, 148
Safety relief valves (PSV), 133, 134–135
Saybolt viscosity meter, 56
Seals
　double mechanical, 80, 81
　mechanical face, 74, 75–82
　single, 78–80
　tandem, 80, 81
Seismic pickup, 62
Service factor, 105, 127
Shockless entrance, 72
Shop driver, 143
Simple harmonic motion, 62
Single seals, 78–80
Sleeve bearing, 83, 104
Sleeve coupling, 123
Sliding vane pumps, 35–37
　diagram, 37
Solar power, 113
Space heaters, 106
Specific gravity, 5
　table, 6
Specific speed, 16–17
Specifications
　pump, 19–20, 42–44
　vendor, 19–20, 154–159
Spur gears, 114, 115
Steam pumps, 30
Steam turbines, 109–111
String test, 143
Stuffing box, 77, 82
Submersible pumps, 29–30
　diagram, 31
Suction
　flooded, 13
　head, 5
　lift, 2, 6, 11, 12, 68–69, 164
　loss, 139
　pressure, 6
　specific speed, 17, 73
Surge control, 134
System curves, 48, 49, 51, 53

T

Tandem seals, 80, 81
Temperature detectors, 137
Tests
　hydrostatic, 143
　mechanical run, 144
　net positive suction head (NPSH), 145
　performance, 143
　string, 143
Thrust bearings, 88, 104
Tilted-pad bearings, 87
Titanium, 99
Total differential head (TDH), 5
Troubleshooting, 151–153
Turbines
　gas, 111–112, 131
　steam, 109–111
Twin screw pumps, 37–38
　diagram, 38

U

Universal joints, 125

V

Vacuum, 2
Valves
　block, 134–135
　control, 128–129
　discharge control, 129–130
　modulating, 128–129
　on-off, 128
　pressure relief, 133–134
　relief, 133
　safety checklist for piping and, 148
　safety relief (PSV), 133, 134–135
Vapor pressure, 2
　table, 3,4
Variable-speed control, 130
Velocity, 62
Vendor data instruction sheet, 140
Vendor drawings, 140
Vendor listings, 165–174
Vendor specifications, 19–20, 154–159

Vertical pump bearings, 88–89
Vertical turbine pumps, 29–30
  diagrams, 30, 31
Vertical volute pumps, 25
Vessel pressure, 69–70
Vibration, 61–65, 153
Vibration limits
  aircraft derivative gas turbines, 64
  centrifugal pumps, 63
  electric motors, 63
  high-speed epicyclic gears, 64
  industrial gas turbines, 65
  low-speed and bevel gears, 64
  reciprocating engines, 65
  rotary pumps, 64
  steam turbines, 64
Vibration monitors, 137

Viscosity
  absolute, 55
  conversion table, 56
  dynamic, 55
  Engler meter, 56
  kinematic, 55
  Redwood meter, 56
  Saybolt meter, 56
  units, 55–56
Viscous liquid performance correction
  charts, 57–58
Voltage, 100–101
Volumetric efficiency, 7

W

Water hammer, 134

Rules of Thumb for Mechanical Engineers

J. Edward Pope, Editor

Experts share tips in the book's 16 chapters packed with design criteria and practical methods that provide quick, accurate solutions to your engineering problems.

1996. 406 pages, figures, tables, charts, appendix, index,
8 3/8" x 10 7/8" paperback.
ISBN 0-88415-790-3
#5790   £42 $79

Reciprocating Compressors Operation and Maintenance

Heinz P. Bloch and John J. Hoefner

For anyone responsible for purchasing, servicing, or operating reciprocating compressors, this text explains how to install, troubleshoot, overhaul, and repair of all types of compressors.

1996. 504 pages, figures, photos, index, appendix. 6" x 9" hardcover.
ISBN 0-88415-525-0
#5525   £55 $85

http://www.gulfpub.com

More Titles for Mechanical Engineers from Gulf Publishing Company

Visit Your Favorite Bookstore

Or order directly from:

Gulf Publishing Company
P.O. Box 2608 • Dept. MW
Houston, Texas 77252-2608
713-520-4444
FAX: 713-525-4647

Send payment plus $9.95 ($11.55 if the order is $75 or more, $15.95 if the order is $100 or more) shipping and handling or credit card information. CA, IL, PA, TX, and WA residents must add sales tax on books and shipping total.

Prices and availability subject to change without notice

Thank you for your order!

Hermetic Pumps

Robert Neumaier, Editor

Hermetic, or sealless, pumps provide leak-free delivery of toxic, radioactive, explosive, and combustible liquids. This volume describes in detail the design, performance, and application of hermetic centrifugal pumps and rotary displacement pumps.

1997. 594 pages, more than 500 figures, tables, charts, index, 7" x 10" hardcover.
ISBN 0-88415-801-2
#5801   £75 $125*

*Not available from Gulf in Germany.

Compressors Selection and Sizing 2nd Edition

Royce N. Brown

An easy-to-read, convenient guide, this practical reference provides the in-depth information you need to understand and properly estimate compressor capabilities and select proper designs.

This second edition is completely updated with new API standards.

1997. 552 pages, figures, photos, appendixes, index, 6" x 9" hardcover.
ISBN 0-88415-164-6
#5164   £60 $80